The Use of Computers for Laboratory Automation

The Use of Computers for Laboratory Automation

S.P. Maj

Adjunkt Professor, Technical University of Denmark, Lyngby, Denmark

ROYAL
SOCIETY OF
CHEMISTRY

ISBN 0-85186-744-8

A catalogue record for this book is available from the British Library

© The Royal Society of Chemistry 1993

Published by The Royal Society of Chemistry,
Thomas Graham House, Science Park, Cambridge CB4 4WF

Typeset by Computape (Pickering) Ltd., Pickering, North Yorkshire
and printed by Hartnolls Ltd, Bodmin

Preface

This book will NOT transform you into a software engineer. It will NOT transform you into a computer engineer. It will NOT even give you the latest computer information. Well, what WILL it do?

Although the number of analysts has remained relatively constant over the years, today laboratories are processing more samples that often display greater analytical complexity. Only increased use of intelligent instrumentation, Laboratory Information Management Systems (LIMSs), and networking, all of which depend on computers, have made this possible. This trend can only continue. Further, increased networking integrates laboratories in the larger manufacturing complex. Doubtless there are few laboratories for which computers do not play an important role in routine operations. This book aims to help scientists understand computers and therefore to use them more efficiently.

Books about computing are numerous, but they are typically written for a wide audience and tend to be product specific. This book was written for scientists by a scientist. Fundamental computer engineering and automation concepts are described in a simple technical language. Sufficient details of the operating principles of computer systems are provided to enable the scientist to read and evaluate technical literature critically and thereby to take an active part in selecting, implementing, using, and maintaining laboratory computer systems. A laboratory computer system is the hardware, software, and procedures needed to perform laboratory functions. Particular attention is given to LIMSs, and to the problems associated with the use, verification, and validation of computers in regulated environments. Further, the emphasis being on principles provides this treatment with the added advantage of being implementation independent and therefore, to a large extent, 'future proof'. The style and content of this book will ensure that it becomes recommended reading for scientists from different disciplines at all professional levels and grades of responsibility.

When this book was started, the most common processor was the Intel 286. By completion, the Intel 486 processor was readily available, and the cost of a personal computer had fallen by 50%—twice the power for half the price! The price information originally included was therefore removed. The reader is encouraged to go directly to the technical sales literature for up to date price details.

What is a computer? Why have they become so important? What developments can we expect next? Five fundamentally important parameters can help us in our considerations: speed; capacity; access time; programmability; and integration scales. Let us consider each in turn.

Speed

A computer works at the speed of its *electronic clock*. To a first approximation each clock 'tick' causes an instruction to be executed. The faster the clock, the faster the execution of instructions. A slow clock speed would be 1 Mhz, *i.e.*, 1 000 000 ticks s^{-1}. Or alternatively each 'tick' takes 1 μs. To put this in perspective, you have almost certainly just blinked. It probably took about 0.1 s. With our computer running at 1 Mhz, 100 000 instructions are executed in the time it takes to blink. Today, a more realistic clock speed is 50 Mhz; tomorrow, clocks will run even faster.

Capacity

You probably have various journals, papers, and reports on your desk. Take a sheet of A4 paper, of which we are wastefully only going to use one side. On that paper there are 200 words. Assuming five letters per word, there are 1000 letters on the page. However, information is also supplied by punctuation marks and spaces. In total then, there are 2000 characters on the page. In your office there is a filing cabinet. Each drawer is 20″ deep and can hold 100 pages per inch. There are five drawers. Each drawer holds 2000 pages. That cabinet stores 20 000 000 characters. An alternative way to view data representation is to imagine that 500 sides of A4 paper (a small book) is the equivalent of 1 000 000 characters.

As we will see in chapters to come, 1K is 1024 and 1M is 1 048 576. But for the time being let us assume that 1K is 1000 and 1M is 1 000 000. The *disk drive* in my cheap desk-top computer is 40 Mcharacters (MB)—twice the capacity of your filing cabinet. Disk capacities of 120 MB are readily available—6 filing cabinets in a device that you could hold on your hand. Computer memories will increase in capacity and decrease in size and cost.

Access Time

Six filing cabinets full of old reports! How long will it take you to find the one you need? In your office there is a memory hierarchy: papers on your desk are readily accessible; papers in your filing cabinet will take a little longer to find. Similarly, there is a memory hierarchy in a computer. Integrated circuit memory chips have nanosecond data *access times*; however, their capacity is usually limited. Disk drives have much larger capacities with millisecond access times. Access times are continually being reduced.

Programmability

No doubt at some time you have adjusted the central heating (or air conditioning) timer so that the heating would come on and go off at the desired times. This unit is programmable and automatic. By using the same mechanism for immersion heating timers, manufacturers can double their sales. Production volumes are substantially increased, and the timers becomes cheaper to produce. Unit cost is inversely proportional to production volume.

The cost of producing the first chip was incalculable. Make two and the cost of each chip is incalculable divided by two. Make four . . . ! Expensive items are used only when necessary. As the cost of these items falls, previously uneconomic applications become economic. Therefore sales increase and the unit cost falls. Now it is possible to buy a microprocessor chip for only a few dollars.

A microprocessor is *programmable*. The chip in a desktop computer could be used in an arcade game, a missile or a washing machine. It is a general-purpose device made application specific by the software. Change the software and the chip will do a different job. As a general-purpose device it can command large production volumes and correspondingly small unit costs.

There are now many readily available software packages (word processing, graphics, *etc.*). This book focuses on the LIMS, a software package written specifically to address the needs of the laboratory.

Integrated Circuits

Some readers may recall looking inside an old radio and seeing large glass elements called thermionic values (or vacuum tubes) glowing red hot. The electrons in such valves travel through a partial vacuum, a phenomenon studied in *gas state electronics*. A valve can be 'on' or 'off'. These valves had the disadvantages of being large, expensive, fragile, inefficient (they get very hot), and unreliable, and were superseded by the transistor in the 1940s. In transistors, electrons travel through semiconductor material. This field is called *solid state electronics*. Transistors are small, cheap, robust, efficient and reliable. Transistors, resistors, and capacitors are connected by means of a printed circuit board (p.c.b.). When individual electronic components can be identified on a p.c.b. it is called a *discrete circuit*. In the 1950s, the transistor fabrication process was extended so that many electronic elements could be included on a single piece of semiconductor material. This was the birth of the *integrated circuit* (IC), for which components do not have to be made separately before connection to the p.c.b. The rule of the discrete transistor was over. Over the years, the scales of integration have increased; they are likely to continue increasing for some time to come (Table 1, Figure 1).

Table 1 *Scales of integration*

Scale of integration	Number of components
Small-scale	10s
Medium-scale	100s
Large-scale	1000s
Very-large-scale	1 000 000s
Ultra-large-scale	1 000 000 plus!

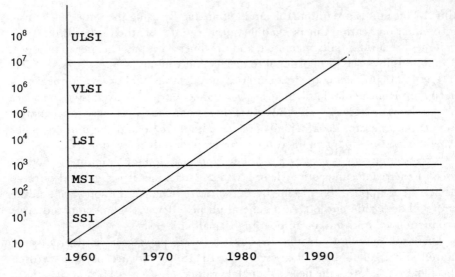

Figure 1 *Integration development*

Contents

Acknowledgements

In gratitude to the people of Denmark and the staff of the Servolaboratoriet of the Technical University of Denmark for some of the best years of my life. Particularly good was 1992: Denmark won the European Cup, we had the best summer in living memory, I survived '5050', and I wrote this book. It was like coming home after having been away for far too long. Special thanks must go to Gudmund Rafn for his friendship and craftsmanship and also Steven Laursen for his '5 minute arguments'. I could not have asked to work with finer engineers or better people.

To Philip and Jan Charlesworth and also Ian and Julie Perkins for their friendship over the years and providing safe mooring during dark and stormy days.

I would like to thank the following for their support and assistance, without which this book would not have been possible. J.L.P.M. Roeling, Laboratory manager, and other staff of Van den Bergh en Jurgens; Graham Walter, Laboratory Microsystems; Dr. Bob Hillhouse, Managing director, and other staff of VG Laboratory Systems; Mr. Martin J. Rosser, Editor, *Pharm. Tech. Int.*; Dr. Bob 'LIMS' McDowall and all of the 'committee'; and Martin Pickering of Harley Systems. I would also like to thank Dr. N.V. Rama Rao, Manager, Hindustan Organic Chemical who is renowned for his contribution to science in India and who was instrumental in arranging my lecture tour of Western India. I am especially grateful to Luc Van de Perre, Marketing Director and other staff of COMPEX nv for their co-operation and support. It was a pleasure and privilege to have worked with COMPEX.

I would also like to thank those who proof-read the manuscript: Olafur Olafsson; Thomas Bech; Soren Kappelgaard; Richard Jonsson; and Ding Hong; and Mike Wetton and Mike Collins for their efforts in proof-reading.

Finally, thanks to Dr. Robert H. Andrews, Senior Commissioning Editor and other staff of the Royal Society of Chemistry for thier patience and support. I hope it has been worth the wait.

CHAPTER 1

Communicating with the Computer—Data Representation

Nods and winks provide various means of communication. Perhaps you can think of others! One important communication method uses the alphanumeric character code set. This set consists of 26 upper-case characters (A to Z), 26 lower-case characters (a to z), 10 numerics (0 to 9), and 32 special characters such as %, *, and space. The importance of this last character cannot be overestimated. Tryreadingthissentencewithoutappropriatespacing. As this character set has to be manipulated, for example, by an automatic typewriter, character codes are needed for control purposes such as carriage return, and form feed. Thirty-four control characters are required bringing the total number of characters in our alphanumeric character code set to 128 (Table 1.1). However, the letters of the alphabet cannot be stored in the computer in the forms we recognize. To see how they are stored we must first consider how information is represented inside a computer.

The basic functional unit of all digital computers is the Binary digIT (BIT). In simple terms this could be a transistor being switched OFF or ON. In this instance, the transistor is acting as a bistable or two-state device, the two states being 0 and 1. Groups of bits (transistors) can be arranged to produce different combinations (Table 1.2).

Do note the pattern of bits shown under Combinations in Table 1.2. The right-hand column reads 01010101, the middle column 00110011 and the left-hand column 00001111. The number of combinations is equal to 2^n, where n is

Table 1.1 *Alphanumeric character set*

Character	Number
A–Z	26
a–z	26
0–9	10
Specials	32
Controls	34
Total	128

1

Table 1.2   *Binary combinations*

Number of bits	Combinations	Number of combinations
1	0	
	1	2
2	00	
	01	
	10	
	11	4
3	000	
	001	
	010	
	011	
	100	
	101	
	110	
	111	8

the number of bits. For 4 bits then, there are 2^4, *i.e.* 16, combinations. The number of combinations doubles for each bit added to the group (Table 1.3).

These different combinations can be allocated different meanings. With 3 bits then, we can have eight different codes. We could, for example, use the allocations of meaning shown in Table 1.4.

Table 1.3   *Binary powers*

Number of bits	2^n
1	2
2	4
3	8
4	16
5	32
6	64
7	128
etc.	

Table 1.4   *Allocated binary meanings*

Binary code	Allocated meaning
000	A
001	B
010	C
011	D
100	E
101	F
110	G
111	H

Table 1.5 *ASCII Table*

BITS 4321 ↓765	000	001	010	011	100	101	110	111
0000	NUL	DLE	SPACE	0	@	P	'	p
0001	SOH	DC1	!	1	A	Q	a	q
0010	STX	DC2	"	2	B	R	b	r
0011	ETX	DC3	#	3	C	S	c	s
0100	EOT	DC4	$	4	D	T	d	t
0101	ENQ	NAK	%	5	E	U	e	u
0110	ACK	SYN	&	6	F	V	f	v
0111	BEL	ETB	'	7	G	W	g	w
1000	BS	CAN	(	8	H	X	h	x
1001	HT	EM	)	9	I	Y	i	y
1010	LF	SUB	*	:	J	Z	j	z
1011	VT	ESC	+	;	K	[	k	{
1100	FF	FS	,	<	L	\	l		
1101	CR	GS	-	=	M	]	m	}
1110	SO	RS	.	>	N	^	n	~
1111	SI	US	/	?	O	_	o	DEL

Note control characters in columns 1 and 2, *e.g.* BEL Bell (audible signal), CR Carriage Return, FF form feed, *etc.*

To communicate the word **HEAD** using our new code we would write 111100000011. The number of codes is rather limited, but still valid. Recall that for our full alphanumeric code we needed 128 codes. This can be provided by 7 bits. To allow free and universal communication there is an agreed 7 bit standard called the American Standard Code for Information Interchange (ASCII), which is shown in Table 1.5.

1.1 PULSE TRAINS

It would be unusual for anyone these days not to have used a computer keyboard. For those who have not, do not worry, they do not often explode when touched! When the key for, say, 'A' is depressed, the associated ASCII binary pattern is generated. This stream of bits, which is called a *pulse train*, is sent down the wire to the computer (Figure 1.1). What is actually transmitted is a series of electrical pulses. Typically, binary 1 is represented by + 5 V and binary 0 by 0 V. Do note that this is not always the case. The pulse train is detected by the computer, manipulated accordingly, and displayed on the screen.

1.2 PARITY

Transmitted electrical signals are, however, subject to electrical interference. Equipment such as heavy machines can produce electromagnetic radiation that can generate electrical pulses in transmission wires. It is possible to transmit one series of pulses but receive a different series. The bit pattern sent may have been 1000001, but a large electromagnetic pulse may have induced a temporary

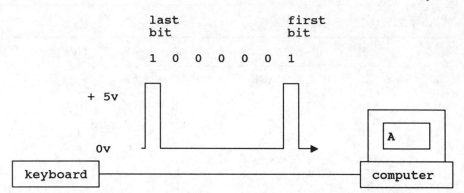

Figure 1.1 *The pulse train*

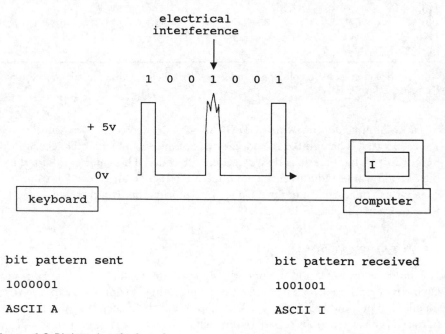

Figure 1.2 *Bit inversion of pulse train*

voltage in the wire between the keyboard and the computer as our pulse train was passing. The result is the production of a voltage spike where there should be none. This is called a *bit inversion*, *i.e.*, binary 0 has changed to binary 1. Let's assume that bit number 4 has been inverted (Figure 1.2). The result is that what is received is 1001001, which is the ASCII code for the letter I. The problem arises because all combinations of the 7 bits are valid. How is the receiving device to know it was the intention not to send I but A?

   To give a degree of protection it is necessary to introduce redundancy. This is done by using 8 bits instead of the original 7 bits. The additional bit is called the *parity bit*. Redundancy allows for the possibility of generating codes that are not

Table 1.6 *ASCII values with even parity*

Value		Parity bit
A	1 0 0 0 0 0 1	0
B	1 0 0 0 0 1 0	0
C	1 0 0 0 0 1 1	1

valid, thus invalidity can be detected. Our basic data unit is now 8 bits long and is known as a *byte*.

There are two types of parity, odd and even. A computer system can use either odd or even parity, but not both. To illustrate this, consider even parity. With even parity, the transmitting device ensures that there is an even number of binary 1s by setting the parity bit accordingly (Table 1.6). The receiving device must be set to the same parity as the transmitting device.

If bit inversion occurs during transmission of the pulse train the parity will change from even to odd. The receiving device employs the same parity as the transmitting device and can therefore detect that an error has occurred. The same principles apply to odd parity.

Table 1.7 shows what happens when the ASCII value A is sent with even parity, but the data is corrupted and the parity changed. The corruption is detected and corrective action is taken. Parity is therefore a simple form of error detection. Although it does not work for double bit inversion, the chances of this occurring are very low for most working environments.

Extended Binary Coded Decimal Interchange Code (EBCDIC) is sometimes called '8 bit ASCII'. As such there are 256 characters in EBCDIC character set. With parity nine bits are needed.

1.3 TRANSMISSION PROTOCOL

On our transmission wire there may be a long stream of bit pulses (Figure 1.3). For the receiving device to decide and interpret this pattern correctly it must know the start and end of each byte, *i.e., byte synchronization*. This method is used when there are long, idle periods between bytes sent at random intervals, such as from a keyboard. The transmission line will be idle for extended periods of time, so the receiver must be informed at the start of each new character. This is achieved by enveloping the data byte between start and stop bits. By convention, the idle line is held high and the first 1 to 0 transition indicates to the receiving

Table 1.7 *Bit inversion of a pulse train with parity*

Bit pattern sent with even parity	Bit pattern received with odd parity (*i.e.*, error)
10000010	10010010
ASCII A	ERROR

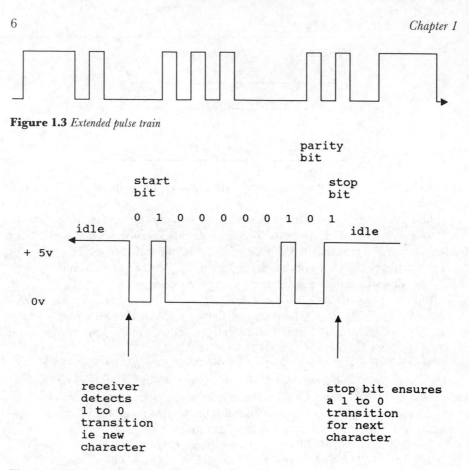

Figure 1.3 *Extended pulse train*

Figure 1.4 *Pulse train with even parity and start/stop bits*

device the start of a new character (Figure 1.4). An idle high line is a legacy of data transmission over the plains of America. Line breaks, which were frequent, could be quickly detected.

1.4 NUMBER SYSTEMS

1.4.1 Decimal Number System

Even though we can now represent our entire alphanumeric character code set by binary codes we cannot do any arithmetic! To be able to perform arithmetic operations such as addition and subtraction we need a number system that uses *positional weighting*. The ASCII code is designed primarily for information interchange and not for the arithmetic manipulation of data. We therefore need another system. The number system we all use in every-day life is the decimal or denary number system. All number systems have a *base* or *radix*. The base dictates the number of characters used to represent quantities in that system. The decimal number system has a base 10 because there are ten digits 0 to 9. All number systems, whatever their base value, have positional weighting. Each digit has a weighting that depends upon its position with respect to the radix point. The weight is determined by a power of the number system base. The denary system has a base power of 10. The positional weight representation of the number 1234, showing the radix point, is

$$10^3 \qquad\quad 10^2 \qquad\quad 10^1 \qquad\quad 10^0$$

$$1000 \qquad\quad 100 \qquad\quad 10 \qquad\quad 1$$

$$1 \qquad\qquad\quad 2 \qquad\qquad\quad 3 \qquad\qquad 4. \quad \text{(radix point)}$$

$$(1 \times 1000) + (2 \times 100) + (3 \times 10) + (4 \times 1)$$

$$1000 \qquad\quad + 200 \qquad\quad + 30 \qquad\quad + 4 \qquad\quad = 1234$$

The value of a number is determined by the sum of products of the specific digit and its associated positional weight. The same applies to fractional numbers but this time the weights have negative powers of 10. The fractional positional weight representation of 123.45 is

$$\text{radix point}$$

$$10^2 \qquad\quad 10^1 \qquad\quad 10^0 \quad . \ 10^{-1} \qquad\qquad 10^{-2}$$

$$100 \qquad\quad 10 \qquad\quad 1 \quad . \ 0.1 \qquad\qquad 0.01$$

$$1 \qquad\qquad 2 \qquad\qquad 3 \quad . \ 4 \qquad\qquad 5$$

$$(1 \times 100) + (2 \times 10) + (3 \times 1) . \ (4 \times 0.1) + (5 \times 0.01)$$

$$100 \qquad\quad + 20 \qquad\quad + 3 \quad . \ 0.4 \qquad\quad + 0.05 \qquad\quad = 123.45$$

The *radix point*, known as the decimal point in the denary system, separates the integer and fractional parts of a number. Note that the *most significant digit* (MSD) has the largest weight and the *least significant digit* (LSD) has the smallest weight.

$$\text{MSD} \qquad\qquad\qquad\qquad \text{LSD}$$

$$1\ 7\ 2\ 3\ 9\ .\ 5\ 4\ 2\ 6$$

This positional weighting makes it simple to perform arithmetic.

1.4.2 Binary Number System

Recall that the binary system has only two states, represented by the digits 0 and 1. As with the decimal number system, each bit (digit) position of a binary number has a positional weight that is determined by a power of the number base system. The binary system has a base power of two. The value of a binary number is again determined by the sum of products. In effect, we have 10 fingers, while the computer has only two! For example, the positional weight representation of binary 1010 is

$$2^3 \qquad\quad 2^2 \qquad\quad 2^1 \qquad\quad 2^0$$

$$8 \qquad\qquad 4 \qquad\qquad 2 \qquad\qquad 1$$

$$1 \qquad\qquad 0 \qquad\qquad 1 \qquad\qquad 0$$

$$(1 \times 8) + (0 \times 4) + (1 \times 2) + (0 \times 1) = \text{denary } 10$$

In general, to convert a binary number to the equivalent denary number, the sum of products is

$$\ldots (b5 \times 16) + (b4 \times 8) + (b3 \times 4) + (b2 \times 2) + (b1 \times 1) \ldots$$

where $b1$ to $b5$ are the specific binary digits.

To convert a denary number to the equivalent binary value, successively divide by 2 to generate remainders, the least significant bit (LSB) being generated first:

$$2 \,\lfloor\, 10 \qquad . \qquad \text{radix point}$$

$$2 \,\lfloor\, 5 \qquad 0 \qquad \text{LSB}$$

$$2 \,\lfloor\, 2 \qquad 1$$

$$2 \,\lfloor\, 1 \qquad 0$$

$$2 \,\lfloor\, 0 \qquad 1 \qquad \text{MSB}$$

With 4 bits there are 16 combinations allowing us to count from 0 to 15 inclusive (Table 1.8). The maximum denary equivalent value of a binary number is given by $2^n - 1$, *i.e.*, zero has to be taken into account (Table 1.9). Increasing the number of bits increases the range of numbers that can be represented. Using 7 bits we have 128 different combinations.

Using positional weighting we can now perform binary arithmetic. We can add two digits A and B to produce the result R and a carry out to the next positional weight. A device that performs this operation is called a *half adder* (Table 1.10). However, with this algorithm we are restricted to adding together only two bits (hence the term half adder). To add groups of bits together we need

Table 1.8 *Four bit system with positional weighting*

$2^3 = 8$	$2^2 = 4$	$2^1 = 2$	$2^0 = 1$	*Denary value*
0	0	0	0	0
0	0	0	1	1
0	0	1	0	2
0	0	1	1	3
0	1	0	0	4
0	1	0	1	5
0	1	1	0	6
0	1	1	1	7
1	0	0	0	8
1	0	0	1	9
1	0	1	0	10
1	0	1	1	11
1	1	0	0	12
1	1	0	1	13
1	1	1	0	14
1	1	1	1	15

Table 1.9 *Maximum denary equivalents*

Number of bits	Maximum denary equivalent value $(2^n - 1)$
1	1
2	3
3	7
4	15
5	31
6	63
7	127
8	255
etc.	

to be able to perform three-bit addition, *i.e.*, *A* plus *B* plus any carry in from the previous positional weight. A device that performs this operation is called a *full adder* (Table 1.11). A 4 bit full adder could perform the following calculation:

$$0101$$
$$+ 0101$$
$$\overline{1010}$$

We are now in the position where we can interpret binary values in different ways! For example, 0000111 may be the ASCII value for BEL or the positionally weighted binary value for denary 7.

The meaning of the bit pattern is context dependent. This is not unusual: for example, 'tak', means 'thank you' in Danish, whereas in Polish it means 'yes'. Each method of interpreting the bit pattern has its advantages and disadvantages. The ASCII code, for example, can be used to represent alphanumeric characters; however, numbers in this system cannot easily be added and are represented inefficiently. The decimal number 123 in ASCII is represented

Table 1.10 *Half adder*

A	B	R	C_{out}
0	0	0	0
0	1	1	0
1	0	1	0
1	1	0	1

	A	0	0	1	1
	+ B	+0	+1	+0	+1
C_{out}	*R*	00	01	01	10

Table 1.11   *Full adder*

A	B	C_{in}	R	C_{out}
0	0	0	0	0
0	0	1	1	0
0	1	0	1	0
0	1	1	0	1
1	0	0	1	0
1	0	1	0	1
1	1	0	0	1
1	1	1	1	1

C_{in}	0	1	0	1	0	1	0	1
A	0	0	0	0	1	1	1	1
B	+0	0	+1	+1	+0	+0	+1	+1
$C_{out}\,R$	00	01	01	10	01	10	10	11

by 24 bits (including parity bits); whereas only 8 bits are needed in the pure binary system.

1.4.3 Other Number-base Systems

Computers can only work in binary. However, binary manipulation is difficult for humans—imagine having to communicate with the computer using only 0s and 1s. In order to simplify this task use is made of other number-base systems, in particular the *octal* (base 8) and *hexadecimal* (hex) (base 16) systems.

We can all count in the denary number system. But let us think for a moment about what we actually do. We count from 0 through to 9, then increase the next positional weight by 1 to 10, and repeat the process. The same principles apply to other number-base systems. The octal system has eight digits 0 to 7. The hexadecimal system has 16 digits, but here we have the problem of what digits to use after 9? In fact we use letters. We therefore count from 0 through to F (Table 1.12)! Each number system has its own associated positional weight (Table 1.13).

As the radices 8 and 16 are powers of 2, these provide convenient representations for numbers in a computer. A string of three bits can take on eight different combinations, hence each 3 bit string can be represented by one octal digit. Similarly, a 4 bit string can be represented by a single hexadecimal digit (Table 1.14). It is very easy to convert a binary number to an octal (or hexadecimal) number. Starting at the binary point and working towards the left, separate the bits into groups of three (or four) and replace the group with the corresponding octal (or hexadecimal) digit. For example,

Table 1.12 *Binary, octal, denary, and hexadecimal counting*

Binary	Octal	Denary	Hexadecimal
0 } coefficients	0 ⎫	0 ⎫	0 ⎫
1	1	1	1
10	2	2	2
11	3 ⎬ coefficients	3	3
100	4	4 ⎬ coefficients	4
101	5	5	5
110	6	6	6
111	7 ⎭	7	7 ⎫ coefficients
1000	10	8	8
1001	11	9 ⎭	9
1010	12	10	A
1011	13	11	B
1100	14	12	C
1101	15	13	D
1110	16	14	E
1111	17	15	F ⎭
10000	20	16	10
10001	21	17	11
etc.			

Table 1.13 *Complete number-base system chart*

			Positional weight			
			Integer			Fractional
Name	Base	Coefficients	MSD	RD*	.	LSD
Binary	2	0, 1	2^1	2^0	.	2^{-1}
			2	1		0.5
Octal	8	0, 1, 2, 3, 4, 5,	8^1	8^0	.	8^{-1}
		6, 7	8	1		0.125
Denary	10	0, 1, 2, 3, 4, 5,	10^1	10^0	.	10^{-1}
		6, 7, 8, 9	10	1		0.1
Hexadecimal	16	0, 1, 2, 3, 4, 5,	16^1	16^0	.	16^{-1}
		6, 7, 8, 9, A,	16	1		0.0625
		B, C, D, E, F				

* RD, Reference digit.

Table 1.14   *Various representations of 3 bit and 4 bit strings*

Decimal	Binary	3 bit string	Octal	4 bit string	Hex
0	0	000	0	0000	0
1	1	001	1	0001	1
2	10	010	2	0010	2
3	11	011	3	0011	3
4	100	100	4	0100	4
5	101	101	5	0101	5
6	110	110	6	0110	6
7	111	111	7	0111	7
8	1000	001 000	10	1000	8
9	1001	001 001	11	1001	9
10	1010	001 010	12	1010	A
11	1011	001 011	13	1011	B
12	1100	001 100	14	1100	C
13	1101	001 101	15	1101	D
14	1110	001 110	16	1110	E
15	1111	001 111	17	1111	F
16	10000	010 000	20	0001 0000	10
17	10001	010 001	21	0001 0001	11
18	10010	010 010	22	0001 0010	12
19	10011	010 011	23	0001 0011	13
20	10100	010 100	24	0001 0100	14

$$101011000110^{binary} = \underset{5}{101} \ \underset{3}{011} \ \underset{0}{000} \ \underset{6}{110}^{binary} = 5306^{octal}$$

$$= \underset{A}{1010} \ \underset{C}{1100} \ \underset{6}{0110}^{binary} = AC6^{hex}$$

$$1011001110101001^{binary} = \underset{1}{001} \ \underset{3}{011} \ \underset{1}{001} \ \underset{6}{110} \ \underset{5}{101} \ \underset{1}{001}^{binary} = 131651^{octal}$$

$$= \underset{B}{1011} \ \underset{3}{0011} \ \underset{A}{1010} \ \underset{9}{1001}^{binary} = B3A9^{hex}$$

In this example note that 0s can be added on the left to make up the correct groupings. Fractional numbers can also be simply represented.

1.5 BINARY CODED DECIMAL

The binary number-base system is called a pure binary code. This is to distinguish it from other types of binary code that are used in computers. Owing to the widespread use of decimal input and output devices, a special form of binary code was developed to be compatible with the decimal system. This is the *binary coded decimal* (BCD) code. The most common type of BCD is known as 8421 BCD and employs a 4 bit binary code to represent decimal digits 0 to 9. This BCD code uses the standard 8421 positional weighting, but there are only 10 valid 4 bit code combinations. The 4 bit binary numbers representing the decimal numbers 10 to 15 are invalid (Table 1.15).

Table 1.15 *BCD chart*

8421 BCD	Decimal
0000	0
0001	1
0010	2
0011	3
0100	4
0101	5
0110	6
0111	7
1000	8
1001	9
0001 0000	10
0001 0001	11
0001 0010	12
0001 0011	13
0001 0100	14
0001 0101	15
0001 0110	16

To represent a decimal number in BCD notation, substitute the corresponding 4 bit code for each decimal digit. For example, decimal 999 would be 1001 1001 1001. This method also applies to decimal numbers with fractions. Certainly, BCD code simplifies bit manipulation, but it is less efficient as it requires more bits to represent a given decimal number. Decimal 15 in pure binary is 1111, but in BCD it is 0001 0101.

1.6 NEGATIVE NUMBERS

In computers, it is necessary to be able to represent signed binary numbers, even though we are restricted to the two binary states, 0 and 1. Of the several methods used for this purpose, we will briefly consider two: sign and magnitude; and two's complement.

1.6.1 Sign and Magnitude

In the everyday manipulation of decimal numbers, we use the sign and magnitude system where a number defines magnitude and a symbol defines whether that number is positive or negative, for instance, -7. The sign and magnitude system can be applied to our binary numbers by using the most significant bit (MSB) to represent the sign $(0 = +, 1 = -)$. For example,

$$00101011 = +43$$
$$10101011 = -43$$

However, the algorithm for arithmetic manipulation is complex and there are two representations of 0 (positive and negative). In a computer, numbers are represented by a *complement number* system.

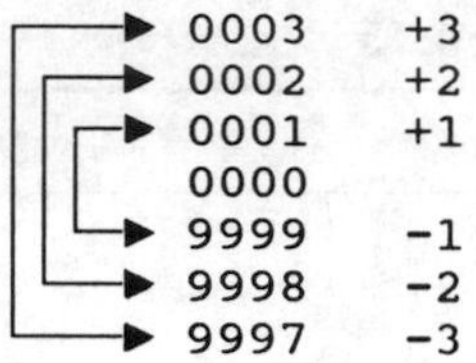

```
    ┌──────► 0003    +3
    │  ┌───► 0002    +2
    │  │ ┌─► 0001    +1
    │  │ │  0000
    │  │ └► 9999     -1
    │  └──► 9998     -2
    └─────► 9997     -3
```

Figure 1.5 *Complements*

1.6.2 Radix Complement Representation

Let us assume that our car has an odometer coupled to its wheels that will record travelling in reverse. Going forward can be considered as positive, going in reverse can be considered as negative. A negative number can be defined as one that when added to its positive aspect will give 0. Positive and negative numbers are complements (Figure 1.5).

During addition and subtraction, if we discard any 'carry' generated we then have a rudimentary number system that does not have an explicit sign that is capable of addition and subtraction.

Sign and magnitude numbers	Radix complement equivalent numbers
+ 3	0003
− 2	9998
+ 1	(1)0001 (discard carry)

Rather than writing out all the complementary numbers, it is easier to complement the individual digits (Table 1.16) and then add 1. For example, we can convert 123 to its negative complement − 123, *i.e.*, 877 (Figure 1.6), and 119 to its negative complement − 119, *i.e.*, 881 (Figure 1.7). Arithmetic (subtraction) may then be performed.

Table 1.16 *Complementary numbers*

Number	Complement
0	9
1	8
2	7
3	6
4	5
5	4
6	3
7	2
8	1
9	0

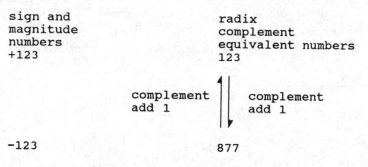

Figure 1.6 *Complements 123 and 877*

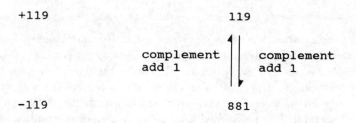

Figure 1.7 *Complements 119 and 881*

Sign and magnitude numbers   Radix complement equivalent numbers

```
+ 123                              123
- 119                              881
```

```
+ 4                          (1)004     (discard carry)
```

The radix complement in the decimal number system is called the *10's complement*. Space does not permit this topic to be investigated further, but you may like to experiment with the numbers and read further on the topic according to your interests.

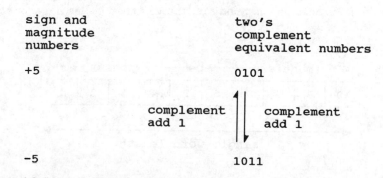

Figure 1.8 *Complements 0101 and 1011*

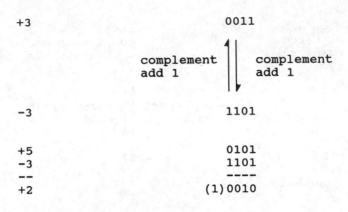

Figure 1.9 *Complements 0011 and 1101*

1.6.2.1 Two's Complement. In the binary system, the radix complement is called the *two's complement.* In this system, if the MSB is 1, the number is negative, when it is 0, the number is positive. As with our 10's complement we can form two's complements (Figure 1.8, 1.9). We have a binary number system without an explicit sign (signless) that can be used to perform arithmetic (addition and subtraction). Subtraction is performed by adding a negative number. Most computers use this form of representation.

1.7 FLOATING POINT

In floating point notation, the operand is split into two fields called the *mantissa* and the *exponent.* A typical arrangement for a 10 bit word length is shown in Figure 1.10. The 'split' is decided at the design stage of the computer. The size of the exponent governs the range of numbers that can be represented, while the size of the mantissa controls the accuracy of numbers within that range. Both the mantissa and exponent are signed binary numbers, usually in two's complement representation. Typically, the mantissa is a fraction and the exponent is an integer.

Floating point notation is the binary equivalent of scientific notation, which operates in base 10. Scientific notation can be used to handle very large or very small numbers.

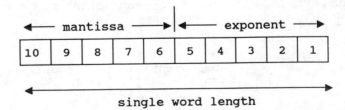

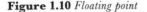

Figure 1.10 *Floating point*

1.8 ERROR DETECTION

Parity will allow error detection, it is simple and easy to implement. However, as we have seen, it cannot detect double-bit inversion. In many applications there can often be bursts of noise, during which more than one bit may be affected.

1.8.1 Block Sum Check

With a block of data (characters), the error detection capabilities of a single parity bit per character may be extended by using an additional set of parity bits computed from the complete block of data (string of characters). This method assigns a parity bit to each character as before, but also generates an extra parity bit for each bit position (column) in the complete block. The resulting set of parity bits for each column is called the *block sum check* (BSC). The rows have a horizontal (or traverse) parity bit and the columns have a vertical parity bit (Figure 1.11, Table 1.17). Double bit errors in a character may be detected. However, this is true only as long as the two-bit errors do not appear in the same column at the same time. The probability of this occurring is not high.

1.8.2 Checksums

The checksum method adds all the values for all the transmitted characters. The resulting checksum is transmitted along with the data. The receiver verifies (checks) the transmission by the sum.

The checksum method is error detection only, and errors must be handled by a retransmission request. Note that in the example we used denary instead of binary values for convenience. Perhaps you are wondering what happens if errors occur to the transmitted checksum? This is a fascinating field and I can recommend many textbooks on the subject, but not this one.

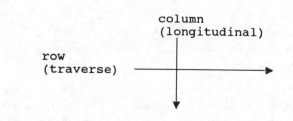

Figure 1.11 *Block Sum Check*, BSC

Table 1.17 *Example of the block sum check*

Character	Bit number*							
	8	7	6	5	4	3	2	1
c	0	1	1	0	0	0	1	1
o	0	1	1	0	1	1	1	1
m	1	1	1	0	1	1	0	1
p	1	1	1	1	0	0	0	0
u	1	1	1	1	0	1	0	1
t	0	1	1	1	0	1	0	0
e	0	1	1	0	0	1	0	1
r	0	1	1	1	0	0	1	0
column parity	0	0	0	0	1	1	1	

* 8 is row parity.

Character	Denary value
c	99
o	111
m	109
p	112
u	117
t	116
e	101
r	114
Checksum	879

1.8.3 Cyclic Redundancy Check

In a *cyclic redundancy check* (CRC), each of the characters in the bit stream is treated as a binary number. This binary number is divided by a predetermined number. The number undergoes the 'modulo 2' operation. This is a mathematical operation that gives the remainder after two numbers have been divided. For example,

$$8 \text{ modulo } 2 = 0 \ (i.e., 8 \div 2 = 4 \text{ with remainder } 0)$$
$$9 \text{ modulo } 2 = 1 \ (i.e., 9 \div 2 = 4 \text{ with remainder } 1)$$

All modulo 2 results are binary values.

Each block of data ends with a special *CRC character*. At the receiving end, the same procedure is performed to find the CRC value. This newly computed CRC value is compared with the transmitted one: if they are the same, there are no errors; if they are different, the block is in error. A binary number can be represented by a polynomial the highest term of which has an exponent one less than the number of bits in the string. For example, the binary number 1010 contains 4 bits that can be represented by a third degree polynomial:

$$
\begin{array}{cccc}
2^3 & 2^2 & 2^1 & 2^0 \\
8 & 4 & 2 & 1 \\
1 & 0 & 1 & 0 \\
\end{array}
$$
$$(1 \times 8) + (0 \times 4) + (1 \times 2) + (0 \times 1)$$

or, more simply, $(1 \times 8) + (1 \times 2)$. This number is divided by the generator polynomial to produce the CRC value. The generator polynomial conforms to agreed standards. For instance, the International Consultant Committee of Telegraph and Telephones (CCITT) generator polynomial is

$$(1 \times 2^{16}) + (1 \times 2^{15}) + (1 \times 2^2) + (1 \times 2^0)$$

which in binary form is 11000000000000101. For the mathematically minded parity checking is equivalent to CRC checking with a generator polynomial of $1 + x$.

It should be noted that the CRC method works extremely well. The CCITT generator polynomial, for example, has the following detection characteristics: all odd numbers of error bits; all possible single error bursts of 16 bits or less in length; 99.9969% of all possible single error bursts of 17 bits; and 99.9984% of all possible longer bursts. But all we have done is to detect the error. Correction is achieved by retransmission.

1.9 STANDARDS

In this chapter, references have been made to standards, such as the ASCII standard. A number of national and international organizations exist to define and enforce standards. This is simpler than it sounds. For a standard to be useful, it should be readily acceptable within current standards and practices, while allowing for anticipated changes due to technical advancement.

1.9.1 Organizations

1.9.1.1 CCITT. The CCITT is part of the International Telecommunications Union (ITU) of the United Nations. As its name implies, the CCITT is largely concerned with telecommunications.

1.9.1.2 International Standards Organization. The International Standards Organization (ISO), which is concerned with a wide variety of applications on an international basis, is becoming progressively important. Its goals include the encouragement of international trade, increased productivity, and quality.

1.9.1.3 Electronics Industry Association. The Electronics Industry Association (EIA) is an American trades organization representing manufacturers of electronic products and interfacing. Standards developed by the EIA often become international standards due, perhaps, to the size and importance of the American market.

1.9.1.4 Institute of Electrical and Electronic Engineers. The Institute of Electrical and Electronic Engineers (IEEE), a professional organization for engineers, is also active in the development of new standards.

1.9.2 Open Systems

Open systems is the attempt to provide vendor-independent systems. As we will see in Chapter 6, Networking, the need to connect a wide variety of manufacturer-dependent equipment in distribution environments led to the development of the Open Systems Interconnection (OSI) communications standard. Open systems attempt to provide maximum portability for applications, data and users, and vendor independence.

These aims have led to the formulation of standards in the areas of communications, operating systems, programming languages, interfacing, user interfaces, *etc*. There are four categories of standards. *Formal standards* are governed by international organizations, for example, the OSI. These standards are developed through an open concensus process that proposes the specification that then defines the standard. *Standard implementations* are based on specifications developed through the open process that result in implementations on different architectures. De facto *standards* are characterized by a very large number of implementations. They are standards due to their prevalence. *Proprietary implementations* are normally owned, controlled, and defined by a single vendor.

CHAPTER 2

Microprocessors

Consider the multiplication operation $12 = 4 \times 3$. To give us a more generally applicable formula for the operation 'multiply' we need a template that defines what we do for any numbers we wish to use, not just for the operands 4 and 3. We therefore need *variables*. Variables are symbolic representations of data, the contents of which can, as the name implies, vary. Variable names are chosen to suit our applications. Consider the general equation for calculating the amount of substance in any volume of a solution: moles = molarity (*i.e.*, moles/litre) × litres, where moles, molarity, and litres are the symbolic representations or addresses of the operands, × is the operation, and = means 'assignment', *i.e.*, multiply the contents of molarity and litres and place the result in moles.

Prior to the operation being completed, values must be assigned to the variables. To execute the operation moles = molarity × litres we must follow the sequence of events shown in Figure 2.1. We have four actions or operations that must be performed in sequence. A set of instructions that perform a specific task is called a *program*. Our aim is to hold the program in a device that will also execute the program—the *stored program concept*. Programs vary in length from a few lines of code to several thousand. Compared with electronic operation, human operation is slow. Recall that our computer clock speeds are measured in Mhz. Our requirements are then: an input device; an output device; a device to process the program; and a device to store the program.

At a simple functional level we therefore have a device that holds our program in memory and executes it, thereby allowing data input and processing, and providing result output (Figure 2.2).

Let us examine in detail the program statement moles = molarity × litres. For this statement to be executed we need the operation code (opcode), ×, the source addresses of the operands, molarity and litres, the destination address of the result, moles, and the address of the next instruction to be performed. In effect, we need the opcode and four addresses, which must be defined somehow in the computer.

2.1 A SIMPLIFIED MICROPROCESSOR UNIT

Our computer model is a simplified microprocessor minus the advanced features (Figure 2.3). It is based on two actual microprocessors, the Rockwell 6502 and

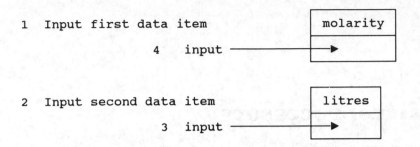

1   Input first data item

         4   input ──────────▶

2   Input second data item

                3   input ──────────▶

3   Perform the operation multiply

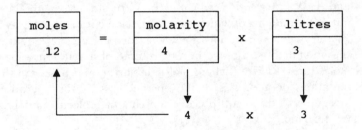

4   Output result

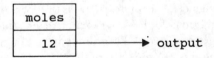

Figure 2.1 *Execution of the operation*

the Motorola 6800. Do note, though for simplicity it is not shown, that all the registers are interconnected allowing data to be transmitted between them.

First, let us identify each of the units in our microcomputer. Note that names vary between manufacturers. The *accumulator* (ACC) is the main working register. It has an implicit address with instructions available for its manipulation. Prior to any operations on data it holds one of the operands, the other is in the *memory buffer register* (MBR). After the operation, the result is stored in the ACC.

The *sequence control register* (SCR) is also known as a *program counter* or the *instruction pointer*. It controls the program sequence, and typically holds the address of the next location to be accessed (rather than the current one).

Arithmetic and logic operations on data are performed by the *arithmetic logic unit* (ALU). There are two inputs, one from the ACC and one from the MBR. The result is placed in the ACC, overwriting the previous contents (Figure 2.4).

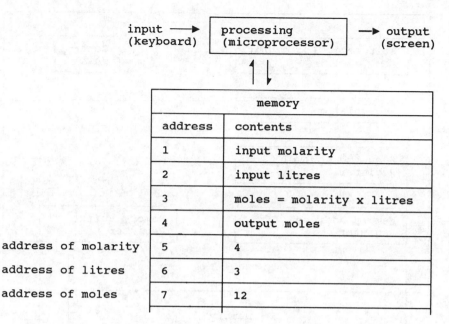

Figure 2.2 *A rudimentary computer*

The control unit sends signals that determine which operation is performed, *e.g.*, ADD, SUBTRACT.

The MBR is a temporary storage location also known as the *instruction register* (IR). It holds data going to and from the data bus, and opcodes prior to decoding. The *memory address register* (MAR) is a temporary storage location that holds the address of the current memory location. The *control unit* (CU) decodes the opcodes and sends out the associated control signals for moving data around.

A series of locations uniquely addressable in any order is called the *memory*. In our case, each unique location can hold 1 byte of data (8 bits). The read/write memory we are using allows data to be written into it or read from it. The *address bus* (AB) from the microprocessor is a series of wires allowing data to be transmitted from one location to another. Our AB and *data bus* (DB) are both 8 bits wide allowing 1 byte of data to be transmitted at a time. An 8 bit address will give access to 265 (denary) locations. The first location is address 00 (hex) and the last address is location FF (hex). Placing an address on the AB allows that specific location of memory to be accessed. At the same time, the microprocessor will send a signal to the memory indicating whether the data is to be read from or written to the memory (Figure 2.5).

A READ signal will cause the contents of the specified location to be placed on the DB and thus transferred to the microprocessor. Do note that the data is only copied onto the DB and is still present in the memory. A WRITE signal causes the data on the data bus to be placed in the location specified by the address on the AB. The name for this type of memory is *random access memory* (RAM). Random access means that all the memory locations are equally accessible.

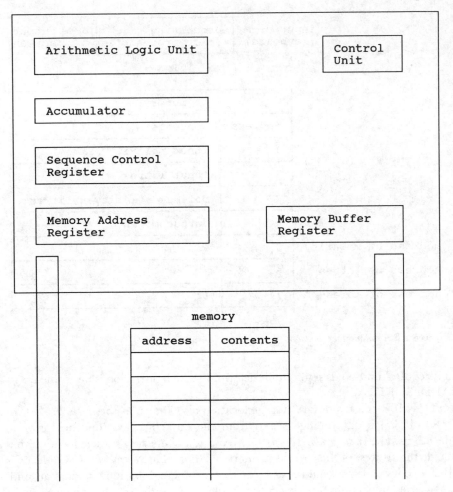

Figure 2.3 *The microprocessor*

However there is another type of memory, read only memory (ROM). This is
also random access, but it does not have the WRITE function.

There is certainly a lot to take in! However, all will become clear when we
breath some life into the theory. This would obviously be best done dynamically
using a series of overheads, but we will use this book just as effectively. I would
strongly encourage the reader to follow the given examples in detail. Your effort
will be rewarded.

The first programmers used binary notation to programme microprocessors,
clearly, a laborious and error-prone task. An improvement is to write the
program in hex. The computer program will then accept hex input and convert
it to binary for us. It is important to realize that, even if the program is in hex,

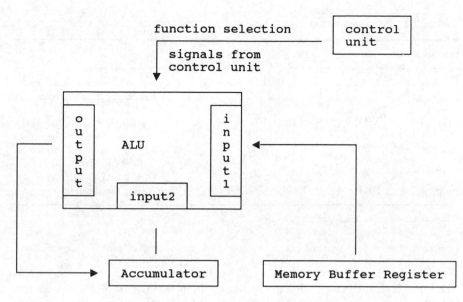

Figure 2.4 *The ALU*

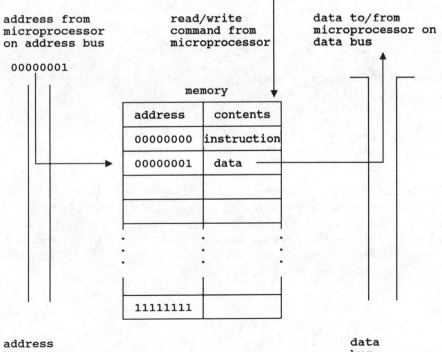

Figure 2.5 *Memory*

Table 2.1　*Some mnemonics*

Operation	Mnemonic	Opcode*	Hex equivalent	Action
Load ACC	LDA	10101001	A9	Load contents of the next memory location into the ACC
Multiply	MULT	01101001	69	Multiply the contents of the next memory location by the current contents of the ACC. Store the result in the ACC
Stop	BRK	00000000	00	Stop the program

* Binary number the computer actually uses.

computers can only work in binary. Therefore the hex-equivalent values we use are solely for our benefit. A further improvement is the use of mnemonics, such as MULT, which the computer converts to the binary equivalent. This is predetermined by the microprocessor manufacturer. All microprocessor operations are defined by specific binary patterns. Table 2.1 lists some operations and their associated binary patterns.

Our program for performing the calculation 4×3 will be

$$\begin{array}{ll} \text{LDA} & 4 \\ \text{MULT} & 3 \\ \text{BRK} & \end{array}$$

However, as the computer can only work in binary, each mnemonic must be replaced by its binary equivalent:

	Binary opcode	Binary data value
LDA 4	10101001	00000100
MULT 3	01101001	00000011
BRK	00000000	

The program consists of three instructions. The first two operands have two parts: an 8 bit opcode followed by an 8 bit operand, the operands being the two numbers to be multiplied together. As we are using an 8 bit computer, the word length is 8 bits, and the microprocessor and memory can only manipulate 8 bit words. As some of our instructions are 16 bits long, they must be broken down into 2 bytes for storage and manipulation. Hence

	Binary opcodes and data values
LDA	10101001
4	00000100
MULT	01101001
3	00000011
BRK	00000000

memory

address	contents
	opcode
	data
	opcode
	data

Figure 2.6 *Immediate addressing mode*

Five bytes of memory are needed. Note that each memory location has two 8 bit binary numbers associated with it—the address and the contents of that address.

2.2 FETCH/EXECUTE OR MACHINE CYCLE

First, a program must be loaded into the memory, then the SCR must be loaded with the address of the first operation in the memory, 00000000. Pressing the GO button, or something similar, causes the program to be executed. The microprocessor then automatically switches between two states *fetch* and *execute*, until the program is completed. This is sometimes called the *machine cycle*. The fetch phase is started first, causing an instruction to be fetched from the memory and decoded by the microprocessor CU. This is followed by the execute phase that carries out the operation previously brought from the memory. These two phases each has a series of steps, rather like the 'tick, tock' of a clock.

Fetch   Execute

Let us now see this in action and in detail. The data can be accessed in different manners or addressing modes. In the immediate addressing mode, the data immediately follows the opcode (Figure 2.6).

2.3 IMMEDIATE ADDRESSING MODE

The fetch/execute cycle of events can be summarized as follows:

	Fetch		Execute
Step		Step	
1	SCR → MAR	1	SCR → MAR
2	SCR + 1	2	SCR + 1
3	MAR → AB → Memory	3	MAR → AB → Memory
4	Memory → DB → MBR	4	Memory → DB → MBR
5	MBR decoded	5	Opcode executed

2.3.1 Fetch Phase, Steps 1–5

By the end of the fetch phase the opcode for LDA has been decoded. Do note that the SCR at this point has changed to the value 00000001 (Figure 2.7). The operation code LDA must now be executed.

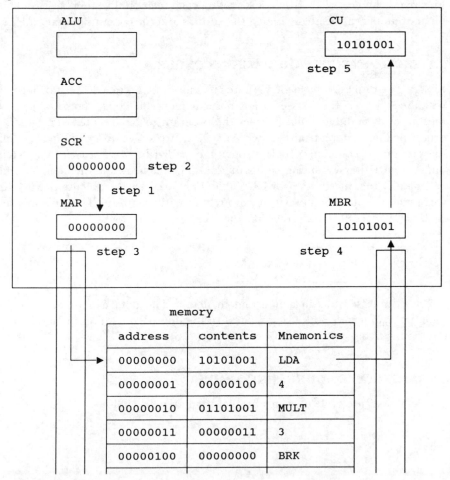

Figure 2.7 *Fetch phase, steps 1–5*

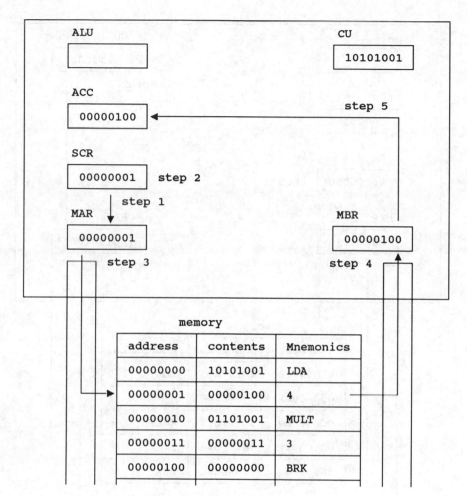

Figure 2.8 *Execute phase, steps 1–5*

2.3.2 Execute Phase, Steps 1–5

The SCR holds the value 00000001, which causes the contents of this location to be accessed. The opcode defines that the contents of this location be loaded into the ACC (Figure 2.8). Note that, after step 2, the SCR has the value 00000010.

2.3.3 Fetch Phase, Steps 1–5

The execute phase has been completed, therefore the fetch phase is started again; however, the SCR points to the address 00000010. In this address is the operation code MULT, which is fetched and decoded (Figure 2.9).

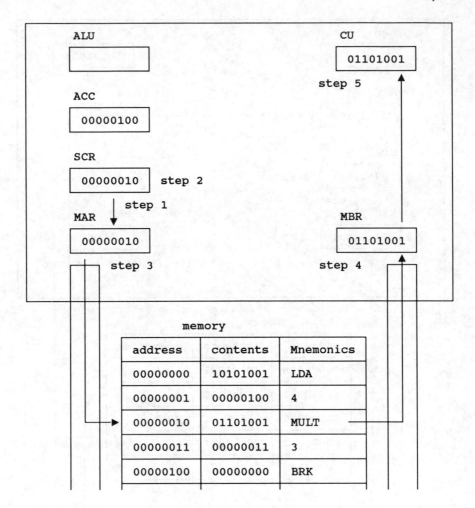

Figure 2.9 *Fetch phase, steps 1–5*

2.3.4 Execute Phase, Steps 1–5

The complete execute phase results in the contents of the MBR and the ACC being presented to the two inputs of the ALU. The two data items are multiplied together and the result is placed in the ACC, overwriting the current contents (Figure 2.10).

2.3.5 Fetch and Execute Phase

The execute phase of the fetch/execute cycle for the instruction BRK terminates the program (Figure 2.11). The complete program requires 5 clock cycles. With a clock speed of 1 Mhz this would give a 5 μs execution time. Obviously the faster the clock, the more quickly the program will be executed.

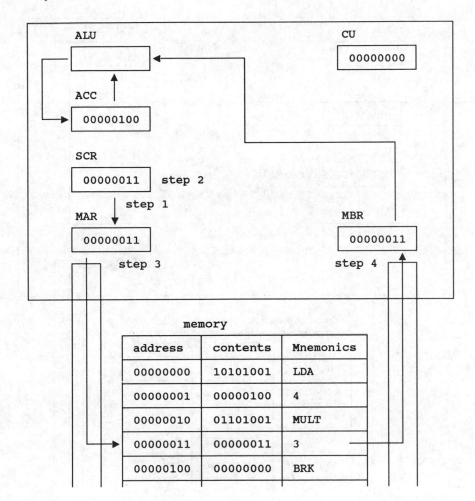

Figure 2.10 *Execute phase, steps 1–5*

It is worth noting that if the SCR was loaded with 0000001 as the start address, the fetch/execute cycle would operate as normal. The CU would therefore try to decode the byte in this location as an operation. However, as 0000001 is not an opcode, the program would halt. It is possible that this byte could be interpreted as a valid instruction. However, it may be something like, 'jump to another part of the memory' (*i.e.*, data and instructions are indistinguishable), and again this would result in a halted program. The meaning of a binary pattern in a byte is context dependent.

2.4 DIRECT ADDRESSING MODE

In immediate addressing mode, the data are embedded in the program. Further, the data are fixed rather like $12 = 4 \times 3$. What we need is a more flexible

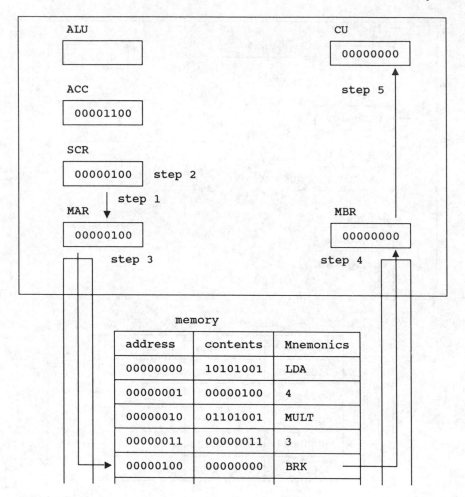

Figure 2.11 *Fetch and execute phase*

approach. This is partly achieved by giving the address of the data after the opcode. The data can therefore be isolated from the program and hence be more easily identified and changed (Figure 2.12). The addressing mode that allows this approach is called the direct addressing mode [from the direct(ions) of where the data are held].

All programs are executed by the fetch/execute cycle, but there will be differences in the number and type of steps required depending on the addressing mode used.

2.4.1 Fetch/Execute Cycle

The difference between the immediate and direct addressing modes lies in step 5 of the execute phase. In the immediate mode, data would be in the MBR at this point, whereas in the direct mode, it is the address of the data that is in the MBR.

memory

address	contents
	opcode
	address of data1
	opcode
	address of data2
address	data1
address	data2

Figure 2.12 *Direct addressing*

Fetch →

← Execute

Step	Step
1 SCR → MAR	1 SCR → MAR
2 SCR + 1	2 SCR + 1
3 MAR → AB → Memory	3 MAR → AB → Memory
4 Memory → DB → MBR	4 Memory → DB → MBR
5 MBR decoded	5 MBR → MAR
	6 MAR → AB → Memory
	7 Memory → DB → MBR
	8 Opcode executed

The MBR contents are copied to the MAR in order to access the data item, step 6. Note that the address is copied into the MAR and not the SCR, which would cause the program sequence to be lost. The extended execution phase is paid for by the need for more clock pulses. The completion of the fetch cycle in both direct and immediate modes results in the decoding of the first opcode LDA. However, as the binary code for LDA is different, this will result in differences in the execute phase.

2.4.2 Execute Phase, Steps 1–5

The completion of the execute steps 1–5 results in the data value being retrieved from memory address 07. As this is an address, it is therefore copied to the MAR from the MBR (Figure 2.13). The value in the SCR, which is now 00000010, has

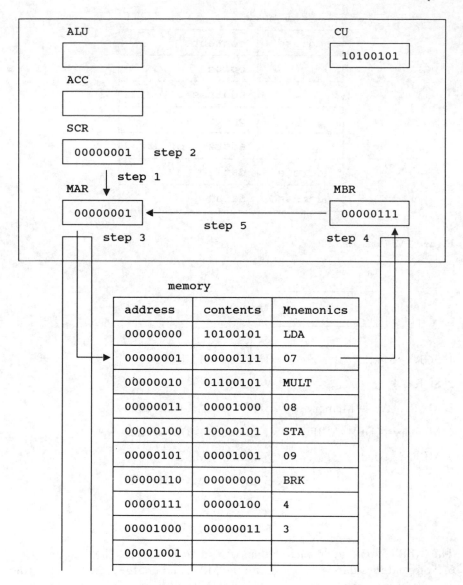

Figure 2.13 *Execute phase (direct mode), steps 1–5*

not been altered. The MAR now holds the address of the data. The MAR contents are placed on the AB and decoded, and the associated data value is loaded into the ACC. This completes the fetch/execute cycle for the first instruction. The contents of address location 7 have been loaded into the ACC (Figure 2.14). As an instructive exercise, you are invited to complete the cycles for the remainder of the instructions.

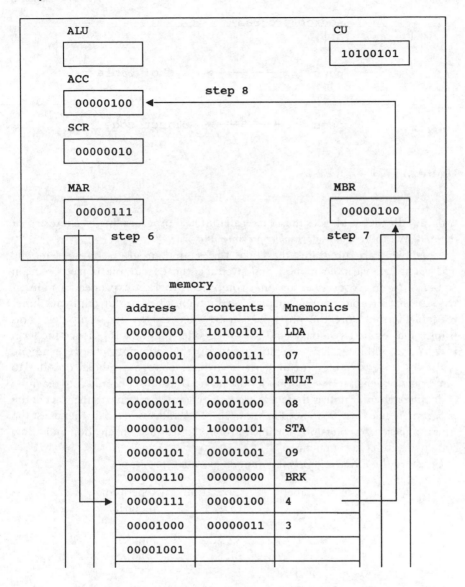

Figure 2.14 *Execute phase (direct mode), steps 6–8*

2.5 TWO-PASS ASSEMBLERS—USE OF VARIABLES

As we are still using addresses rather than variable names, there is more work to be done. Programs in simple assembler are difficult to change, because absolute addresses are used. If a new line is inserted into the program all other lines must be modified. Program addresses must be allocated with due regard to opcode and operand size. With direct addressing, we isolate the data from the program, but we do not know where exactly to place the data until we have written the

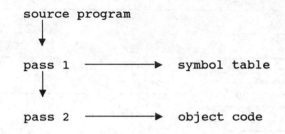

Figure 2.15 *Two pass assembler*

program. Branch addresses must be recalculated if program lines are inserted or deleted. We therefore use *assembler programming*.

Assemblers are programs that take the source program (using symbolic addresses, *i.e.*, variable names) and produce the executable object program (binary). Source code is an original computer program expressed in human-readable form (programming language), which must be translated into machine-readable form before the program can be executed by the computer. This translation requires two passes of the source code (Figure 2.15), hence the term two-pass assembler. These assemblers can do the following: change symbolic addresses (*i.e.*, variables) to numeric addresses; change symbolic opcodes to binary equivalents; reserve storage for data; and translate constants into machine code. Of particular note is that addresses are generated relative to the start of the program, hence there is no need to know absolute addresses. To further assist the programmer, assemblers provide line numbers for error identification and allow statements to have comments.

Consider the following assembly language source program:

Line	Label	Operation	Operand
1		LDA	molarity
2		MULT	litres
3		STA	moles
4		BRK	
5	molarity	4	
6	litres	3	
7	moles		

The associated instruction set for the microprocessor is

Symbolic code	Machine code	Length	Operands
LDA	A5	2	1
MULT	65	2	1
STA	85	2	1
BRK	00	1	0

Pass 1 is executed as follows. First, the source text is scanned to determine the relative addresses from the start of the program.

Table 2.2 *Pass 1 generated symbols*

Symbol	Address
Molarity	7
Litres	8
Moles	9

Line	Address	Label	Operation	Operand
1	0		LDA	molarity
2	2		MULT	litres
3	4		STA	moles
4	6		BRK	
5	7	molarity	4	
6	8	litres	3	
7	9	moles		

This enables the assembler on pass 1 to generate the symbol table (Table 2.2) that lists the addresses of the variables. This is necessary as *forward references* are made to variables.

The assembler program can then construct the following program from these direct addresses:

Address	Contents
0	LDA
1	7
2	MULT
3	8
4	STA
5	9
6	BRK
7	4
8	3
9	

During pass 2 the mnemonic codes can be substituted by their binary equivalents and the addresses of the variables from the symbol table (Table 2.3).

Table 2.3 *Pass 2 code generation*

Address	Machine code
0	A5
1	07
2	65
3	08
4	85
5	09
6	00

2.6 LOW-, MEDIUM-, AND HIGH-LEVEL LANGUAGES

The microprocessor can be programmed laboriously in binary, opening the way for errors. As we have seen, one improvement is to write the program in hex. We then have in the computer a program that will accept hex input and convert it to binary. The second improvement was to use symbolic names or mnemonics for each instruction, opcodes. Now we can type meaningful programs like LDA OF, but we must give absolute addresses. Hence the program and data can be loaded into only one place in memory, and that place must be known in advance. Any changes in the program will change the addresses of all subsequent lines. So it is much better, therefore, to use symbolic addresses. *Assembly programs* are programs containing symbolic addresses. They are translated into machine language by an *assembler* program. We can now write things like LDA molarity. These programs are *machine dependent* and will not work on different microprocessors; being machine specific they are called low-level languages. For example, the opcode LDA works on the Rockwell 6502, but not on the Motorola 6800, which has two ACCs. Change or upgrade your computer and you will have to rewrite all your programs! To overcome this problem, high-level languages such as Pascal were developed. A Pascal program should work on any computer. Typically, a high-level language program has the following structure:

> Name of program
> Variable declarations
> Statements to be executed

More specifically, we have the following:

> Name of program
> Molarity, litres, moles
> BEGIN
> Input molarity
> Input litres
> Moles = molarity × litres
> Output moles
> END

In the programming language Pascal this would be written as

```
PROGRAM Moles (Input, Output);
VAR moles, litres, molarity : INTEGER;
BEGIN
    WRITE('Please enter moles and litres');
    READ(moles);
    READ(litres);
    moles = molarity * litres;
    WRITE(moles);
END.
```

Compiling is the process of converting a high-level language program from human-readable to machine-executable form. After compilation the program

Table 2.4   *Expressing the code, moles = molarity * litres*

| | | Contents | | |
Memory Address	Full assembly code with symbolic addresses	Symbolic code with hex addresses	Hex code with hex addresses	Binary code with binary addresses
00000000	LDA	LDA	A5	10100101
00000001	molarity	07	07	00000111
00000010	MULT	MULT	65	01100101
00000011	litres	08	08	00001000
00000100	STA	STA	85	10000101
00000101	moles	09	09	00001001
00000110	BRK	BRK	00	00000000
00000111	4	4	4	00000100
00001000	3	3	3	00000011
00001001	12	12	12	00001100

may be run. It will prompt the user to input two values by displaying on the screen, 'Please enter moles and litres'. The data is entered from the keyboard. The amount of substance is calculated from the data values supplied and the result is output to the screen. The Pascal line of code moles = molarity * litres can be written in a variety of ways (Table 2.4): full assembly with symbolic addresses; symbolic code with hex addresses; hex opcodes with hex addresses; and binary code with binary addresses, which is how the computer actually works.

It is important to realize that in the Pascal program, and in other high-level languages, the variables have to be *declared*. This is achieved by the line

VAR moles, litres, molarity : INTEGER

This is an instruction to the computer that defines the *data type* of the variable or variables. In our example, the variables hold integers, therefore they CANNOT be used to store floating point numbers or characters. The storage requirements of the data types are often different. As we have seen, floating point numbers need more bytes for storage than do integer numbers. Even if the storage requirements were the same, the computer would have to be informed of the valid operations that could be performed on the data. Recall that normal arithmetic operations cannot be performed on characters.

2.7 THE 6502 MICROPROCESSOR

Our microprocessor is modelled on the Motorola 6800 and the Rockwell 6502. The architectures of processors are, however, more complex than our simplified model, as illustrated by the pin-out diagram of the 6502 processor (Figure 2.16).

The RESET stops the processor in its current action and gives control to the monitor (a simple operating system). The processor must be able to interact efficiently with external devices. These can inform the processor that they need attention by the two interrupt lines. The interrupt request IRQ may or may not

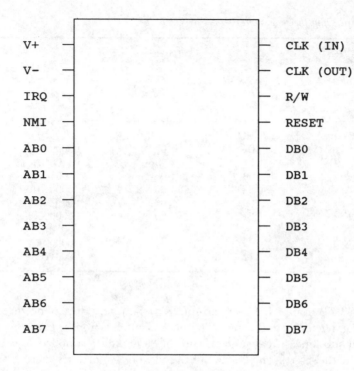

The pins can be identified as follows :

AB0 to AB7 : address bus

DB0 to DB7 : data bus

V+, V- : power lines

RESET : give control of the microprocessor to the monitor

IRQ : Interrupt request

NMI : Non maskable interrupt

CLK (IN) : Connections for external clock (1 Mhz)

CLK (OUT) : Provides clock signal to other devices

R/W : Read/Write

Figure 2.16 *The 6502 microprocessor*

be given attention, and NMI is an interrupt request that cannot be ignored. An external clock provides the clock pulses. The R/W line signals when the microprocessor wants to write to or read from a memory or another device (Figure 2.17). We will read more about the operation of these pins in Chapter 5, Input and Output Interfacing—the Outside World.

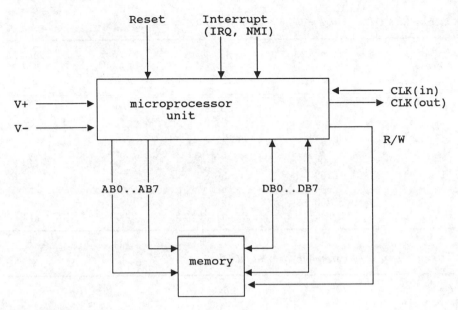

Figure 2.17 *Pin functions*

2.8 MICROPROCESSOR PERFORMANCE

The performance of a microprocessor will depend on various factors including clock speed, MAR size and AB width, DB and ACC width, and number of instructions available for use.

The faster the clock, the more quickly the programs can be executed. The wider the MAR and AB, the larger the number of memory locations that can be addressed. Note that often the MAR and AB widths are not equal. Our 8 bit AB will allow us to access 2^8, *i.e.*, 256 (denary), unique memory locations. This is rather limited, and modern computers have much larger ABs, typically bit length multiples of eight (*e.g.*, 16 bit) (Table 2.5). As levels of integration increase along with demands for access to even larger data volumes, ABs and the associated DBs are increasing in size. It is not untypical to have 32 bit ABs. With 16 bit AB, we can access 64K addresses, where K = 1024. A 24 bit AB will allow access to 16M addresses, where M = 1024 × 1024. A 32 bit AB accesses 4G addresses, where G = 1024 × 1024 × 1024. The pattern is as follows (Table 2.6):

$$1N \text{ AB will access } N^2K \text{ addresses}$$
$$2N \text{ AB will access } N^2M \text{ addresses}$$
$$3N \text{ AB will access } N^2G \text{ addresses}$$

where N, $2N$, and $3N$ are bus widths.

With a DB and ACC width of 8 bits we can only handle denary values up to 255. To overcome this problem, with an 8 bit machine it is necessary to perform multibyte arithmetic, *i.e.*, 2 bytes for double precision, 3 bytes for triple precision, *etc.* In double precision arithmetic, the low-order bytes must be added first with

Table 2.5 *Memory size with up to 16 bits*

Number of address lines	Number of memory locations	Memory size/ K	Address low hex	Address high hex
8	256	0.25	00	FF
9	512	0.5		
10	1024	1		
11	2048	2		
12	4096	4	000	FFF
13	8192	8		
14	16384	16		
15	32768	32		
16	65536	64	0000	FFFF

Table 2.6 *Memory size with up to 40 bits*

Bus width/ bits	Memory*	Bus width/ bits	Memory*/K	Bus width/ bits	Memory*/M	Bus width/ bits	Memory*/G
1	2	11	2	21	2	31	2
2	4	12	4	22	4	32	4
3	8	13	8	23	8	33	8
4	16	14	16	24	16	34	16
5	32	15	32	25	32	35	32
6	64	16	64	26	64	36	64
7	128	17	128	27	128	37	128
8	256	18	256	28	256	38	256
9	512	19	512	29	512	39	512
10	1024 = 1K	20	1024K = 1M	30	1024M = 1G	40	1024G = 1T

* K = 1024;   M = 1024K = 1024 × 1024;   G = 1024M = 1024 × 1024 × 1024;   and T = 1024G = 1024 × 1024 × 1024 × 1024.

any carry over passed to the addition of the high-order bytes. Obviously much more memory and computation time is needed. Processors with 16 bit data highways can perform 16 bit arithmetic with one operation.

The range of instructions provided for use on a microprocessor is called the *instruction set*. Microprocessors are progressively becoming more complex, with increasing numbers internal registers and instructions. Data held in internal registers can be accessed more quickly.

These trends can be demonstrated by following the evolution of Intel microprocessors (Table 2.7).

2.9 MICROPROCESSORS, PERSONAL COMPUTERS, MINICOMPUTERS

It used to be possible to classify computers as micro, mini or Mainframe according to data width. A personal computer would have been 8 to 16 bits, a

Table 2.7 *Specifications of Intel microprocessors*

Model*	Year introduced	Number of instructions	Register width/bit	AB width/ bit	Addressable memory	DB width/ bit	Clock speed/ MHz
8088	1978	111	8	16	64 Kbyte	8	1
8086	1978	133	16	20	1 Mbyte	16	10
80286	1982	hundreds	16	24	16 Mbyte† 1 Gbyte‡	16	10 plus
80386	1985	hundreds	32	32	4 Gbyte† 64 Tbyte†	32	40 plus

* The Intel Pentium (80586) has 64 bit capability.
† Physical memory (see Chapters 3 and 4).
‡ Virtual memory (see Chapters 3 and 4).

minicomputer 16 to 32 bits, and a Mainframe more than 32 bits. Today you can have microprocessors with 32 plus bits on your desktop personal computer (PC). This rate of development demands looser definitions (Table 2.8).

As we have seen, a microprocessor is a complex electronic circuit that performs arithmetic, logic, and control operations. It is a single integrated circuit (IC), *e.g.*, the Rockwell 6502, consisting of many thousands of transistors on a single silicon chip about 0.125″ square. The chip is situated in a package with leads for connection and ease of handling.

A microcomputer (*i.e.*, the PC) contains a microprocessor, but also holds other support circuits, such as memory devices for storing programs and data, interface adapters for allowing communication with the outside world, and a clock. Typically, the chips are placed on a *printed circuit board* (p.c.b.) that has all the wiring that connects the ICs. These data highways consist of groups of parallel conductors called buses over which bits can pass. The p.c.b. that holds the microprocessor and associated chips is often referred to as the *motherboard*. Other p.c.b.s may be plugged into the motherboard.

Input data can come from a variety of devices such as keyboards, other computers, and instruments. Output data can also be sent to a variety of devices such as visual display units (VDUs), and printers. These are typically called input/output or I/O devices and are connected to the computer by I/O ports (Figure 2.18).

For data movement, the reference point is the microcomputer. Information passing into the microcomputer is called input data. Information transmitted to the outside world is called output data. Information passes to and from the microcomputer via an I/O port.

There are many IBM and IBM-compatible computers on the market today. The Industry Standard Architecture (ISA) was originally an 8 bit DB. With the advent of the Advanced Technology (AT) model this was extended to 16 bits. The Enhanced Industry Standard Architecture (EISA) Bus is a 32 bit specification. The Micro Channel Architecture (MCA) is also a 32 bit DB, but the architecture is owned by IBM. Manufacturers wishing to include the MCA in a PC can only do so under licence. The EISA standard is a consortium of vendors

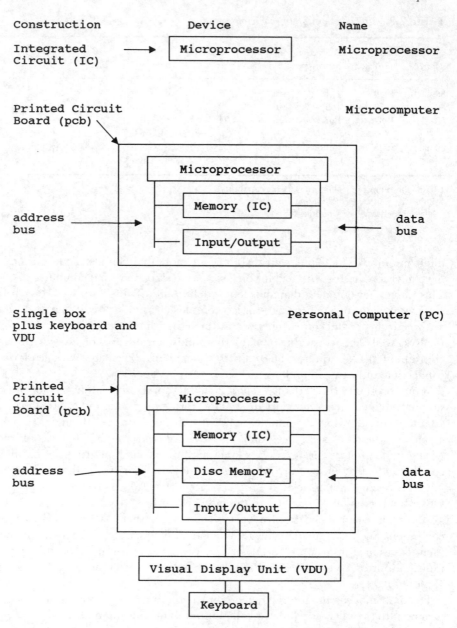

Figure 2.18 *Microprocessors, PCB's, and PC's*

Table 2.8 *Computer classification*

Computer type	Construction	Number of users	Environment	Example	Cost/£
Micro	Single p.c.b. in a single slim box, plus a keyboard and screen	1	Ambient	IBM PCs PS/2 range	100s
Mini	More than one p.c.b. in one or more boxes, plus many keyboards and screens	10s	Ambient	HP 9000	1000s
Mainframe	Large cabinet(s), plus many keyboards and screens	Many!	Air-conditioned	Cray	10 000s +

with their own bus specification to prevent (excessive) royalty payment to IBM. For hard- and software there are two basic systems with subcategories: PC and PC extended (XT) systems—8086 central processing unit (CPU) (8 bit ISA) and 8088 CPU (8 bit ISA); PC AT systems—80286 CPU (16 bit ISA) and 80386 CPU (16 bit ISA); and PS/2 MCA bus—80286 CPU (16 bit MCA), 80386 CPU (32 bit MCA), and 80486 CPU (32 bit MCA); and EISA bus—80386 CPU (32 bit EISA).

2.10 COMPUTER POWER AND PERFORMANCE

To a first approximation, the power of a computer depends upon factors such as clock speed, register size, and the instruction set. But, is an Intel 386 based computer running at 33 MHz better than an Intel 486 computer running at 25 MHz? Or is a Motorola 68030 based computer better than an Intel 486 based computer? It is important to realize that the computer hardware system also consists of primary (with and without cache) and secondary memory. A slow disk drive would obviously inhibit the performance of a powerful microprocessor. Furthermore, in the final analysis the computer will be subjected to different workloads. The efficiency of the computer in dealing with workloads depends not only on the hardware but also on the operating system and the application packages used. It is essential therefore to consider the complete computer system. To address these often-asked questions we must consider the use of the benchmarks, performance testing, and tuning (Figure 2.19).

Benchmarks are a means of comparing a number of possible computer systems in order to make a quantifiable, manufacturer-independent selection. However, it is important to stress that a benchmark is a static test that is used before a system is installed. There are inherent dangers in extrapolating to actual working

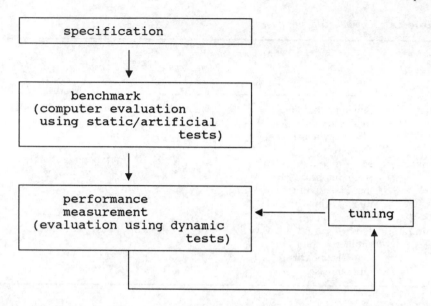

Figure 2.19 *Stages in the selection, performance measurement, and tuning of a computer*

environments. A wide range of benchmarks of varying complexity can be used—Landmark speed, Norton SI, Whetstones, Dhrystones, *etc.* For example, the Gibson Mix consists of timing the program instructions of each computer and then weighting them by percentages that represent their frequency of use. On-line transaction processing (OLTP) is a benchmark for evaluating computer systems and, as the name implies, takes into account on-line data processing requirements. For some applications the demands on the system may be 'peaky', and considerations include: interactive response time; interactive response time at peak load; and batch throughput.

After a computer has been installed it is important to conduct a performance evaluation. Indeed this may be essential if the response times are unacceptable. Furthermore, computer systems are not static and within the lifetime of the computer system progressively greater demands may be made upon it. Perform-ance measurement is the evaluation of a computer system under operating conditions in order to identify inefficiencies quantifiably and to correct them accordingly thereby improving system efficiency. Performance measurement is therefore dynamic and directly relevant to the working environment. The main objectives include: workload analysis (*i.e.*, even or peaky); identification of system bottlenecks (*e.g.*, disk throughput); identification of programs and/or users that place loads on the system sufficient to degrade its performance; provision of a means of evaluating new hard- and software; and generation of benchmarks relevant to the working environment that may be used to evaluate new computers.

The two main methods of performance measurement are hard- and software monitors. Hardware monitors are electronic devices that are connected into the computer. As such they place no overhead on system performance, but tend to be expensive. Software monitors are programs that run at the same time as all the other software and therefore place an overhead on the system degrading its performance! When a system can be measured, it is then possible to improve its performance by tuning.

2.11 REDUCED INSTRUCTION SET COMPUTERS

In Table 1 we noted that the number of instructions in the instruction set of the microprocessor has progressively increased, the primary motivation being to improve microprocessor performance. For example, simple microprocessors do not have an opcode for multiply. Instead, a small assembly language program must be written to perform the multiply operation. Increasing the instruction set should therefore result in smaller and faster programs. The computers based on this design strategy are called *complex instruction set computers* (CISCs).

Studies in the late 1970s demonstrated that high-level languages were best supported by optimizing the performance of the most time-consuming features of programs written in these languages. These studies demonstrated the desirability of the following design features: a limited and simple instruction set; a large number of general purpose registers, one instruction per cycle; register to register operations; simple address modes; and simple instruction formats.

Reduced instruction set computers (RISCs) were designed along these principles. One example is the 64 bit RISC Alpha chip from Digital, which is claimed to run equally well on any operating system. High-performance RISC architecture is typically used in minicomputers and workstations with a progression towards PCs.

2.12 PREVENTATIVE MAINTENANCE

Computer systems (hardware, interfaces, instrumentation, *etc.*) need regular maintenance. Preventative maintenance is the principle method of ensuring trouble-free operation with respect to component failure, data loss, *etc.* There are two main types of preventative maintenance (PM), active and passive. Active PM is maintenance in anticipation of component failure. Passive PM comprises the measures taken to protect the system from the environment, *e.g.*, power-surge protection devices. The frequency of active PM depends upon the equipment and the working environment. A computer system located in the laboratory of a steel mill will be in a considerably dirtier environment than one in a biochemistry laboratory. Take advice from your supplier.

One important aspect of PM is the regular and thorough cleaning of your system. Many computers use forced-air cooling. An exhaust fan is used to draw outside air through cooling slots in the chassis. This air is not filtered, and consequently dust and airborne contaminants accumulate in the computer. Depending on the composition of contaminants, the effects vary. At a minimum, dust will act as a thermal insulator preventing proper cooling of the ICs. In extreme cases the airborne contaminants will lead to short circuits between components and corrosion of the electrical connections. The drawn air also flows in through the floppy disk port. Hard disks, however, are hermetically sealed. In industrial applications it is often necessary for the action of the computer fan to be reversed. Filtered, dry air is blown into the computer box thereby creating a slightly higher pressure that exhausts through the chassis. This prevents airborne contaminants entering the computer. This is one of the many steps taken to make industrial computers more robust. In common with other measures such as thermal insulation, and p.c.b. locks, it adds to the cost. Integrated circuits do work loose. However, reseating socketed chips, cleaning p.c.b.s, and reformatting hard disks should be left to the experts.

2.13 OPERATING ENVIRONMENT

Factors affecting the operating environment are now considered. As mentioned in the previous section, airborne contaminants can have a detrimental effect. Computer systems should therefore be protected from airborne contaminants whenever possible.

Computers have a recommended working temperature range. Typically, two sets of figures are provided, the operating maximum and minimum. Keep computers away from direct sunlight and windows that will cause large temperature variations. Large temperature changes will physically stress computers and equipment owing to thermal expansion and contraction. This may eventually cause problems, as, for example, excessive chip creep. Mainframe computers must be maintained in environmentally controlled atmospheres.

One of the largest temperature variations is caused by system start up—power cycling. Reducing the number of times you switch on your computer will increase its reliability. Just think, it is when you turn light bulbs on that they fail! It is not unusual for the computer to be running, even if not used, for 24 h per day. One compromise is to switch on your system only once a day—your hard disk will not suffer.

Static electricity is mainly a problem in dry climates. The problem will be exacerbated if the chassis is not properly grounded.

Electromagnetic interference (EMI), more typically called radio frequency interference (RFI), even if it does not cause radio interference, is caused by radio transmitters, portable telephones, arc welding equipment, large electric motors, *etc*. Any wire can act as an aerial with voltages being induced in it by electromagnetic fields. Reorientation can help reduce the effects of interference due to the directional nature of radio waves. Alternatively, it is possible to buy shielded cables for wires that go to and from the chassis. Do note that your computer is also a source of RFI, and therefore computer equipment must be certified accordingly. This topic will be considered in more detail in the chapters to follow.

A recent survey in the US claimed that 50% of all computer failures are power related and that 85% of companies that suffer a major computer catastrophe go out of business within 18 months. The power supply of a computer is shared by many other users. If other users include heavy-duty motors these might cause voltage variations or even the transmission of very high voltage spikes, called transients, along the power lines. A brown-out is when a power line falls below 80% of its nominal voltage. This typically occurs when there is a surge in demand on the power grid. A spike is an over-voltage lasting between 1 and 100 ms at an amplitude of 100% or more of the line voltage. In the UK, there is a guaranteed supply of 240 V at 50 Hz with the standard allowing a voltage–frequency variation of $+/-6\%$. For many electrical appliances the quality of supply is satisfactory—at worst the lights may flicker. Computers are much more sensitive, the result of poor supply quality being problems that range from data corruption to equipment damage. Buying a good quality computer system will help ensure some degree of protection by providing over-voltage/current protection and power conditioning; however, these do not fully address the problems of guaranteed supply, and excessive noise on the power lines. Consequently there is a progressive trend towards further protection in the form of either standby power supplies (SPS) or uninterruptible power supplies (UPS).

2.13.1 Standby Power Supplies

Standby power supplies allow graceful shut-down. They are designed to provide enough power for data to be saved, after the mains have failed. They detect any deviations in the power supply line and switch to a standby battery and power invertor when necessary. A power invertor converts the low d.c. battery power to high voltage a.c. Uninterruptible power supplies must switch between the power line and battery supply.

2.13.2 Uninterruptible Power Supplies

Uninterruptible power supplies allow the user to continue working as normal for extended periods of time, and provide surge protection. There are two types of UPS, on-line and off-line. Off-line UPS are the cheaper and less sophisticated. These are not strictly UPS but rather a standby power filter that simply routes the mains through itself and provides filtering as it charges its batteries. If there is a break in the power supply the batteries can then take over. Power Supplies to PCs contain filtering capacitors that take about 5 to 10 ms to discharge, giving the UPS time to switch in. Within about 3 to 4 ms of the mains power being cut, a sensing device switches the output to the batteries and an invertor. Off-line systems are cheaper, as they do not run continuously. Perhaps the main problem with off-line systems is that you do not know if the system is going to work when it has to be switched on. In contrast, on-line UPS are on all the time. The mains power is converted to d.c. power batteries. The d.c. power then enters an invertor that converts it back to a.c. The effect is that the PC is running off the batteries the whole time. If the power fails there is no transfer time or fluctuation. The main disadvantage is cost.

In a network, the file server is the single most important item and the use of UPS, which give a 'prayer period', is invaluable. Failure of the server would halt the whole system, putting a lot of data at risk. This is especially true where diskless workstations are used. Uninterrupted power supplies can be considered as essential equipment for any application-critical computer system. We will meet file servers and workstations in Chapter 6, Networking.

Factors taken into account during site preparations should therefore include the provision of separate circuits for the computer systems and a good ground connection, and avoidance of the use of extension cords (noise increases with resistance, which is a function of cable length). In particular, do not site your computer where power tools and vacuum cleaners are frequently used, or near air conditioning units, *etc.*

2.14 WARRANTY AND SERVICE CONTRACTS

Typically, many suppliers offer an on- or off-site 1 year warranty that may be extended. The service contracts may be offered by the manufacturer, an authorized dealer, or by a third party. Increased levels of service are available at a cost, typical arrangements being a 1, 2, 3 or 4 h on-site response, the same or next day on-site response, or a body bag or carry-in service (you take the computer to the dealer!).

CHAPTER 3

Memory Devices

Memory is used to store data and data processing instructions. The demand placed upon memory means that no one technology is suitable for all our requirements. To be of most use, memory devices should have the following characteristics: cheapness; non-volatility (*i.e.*, when the power is turned off the data should not be lost); erasability (*i.e.*, the possibility of storing and deleting data, and reusing the location); random access (*i.e.*, all information is equally accessible); and short access time. The storage medium we have been using so far fails on the first two characteristics! It is overly expensive to have, say, 200 Mbytes of RAM, and this type of memory is volatile—turn the power off and your data are lost.

3.1 MEMORY HIERARCHY

Computer storage consists of a mixture of technologies that are placed in a *memory hierarchy* (Figure 3.1). To provide permanent mass storage (hundreds of mega-bytes) expensive and volatile integrated circuit (IC) RAM chips are used together

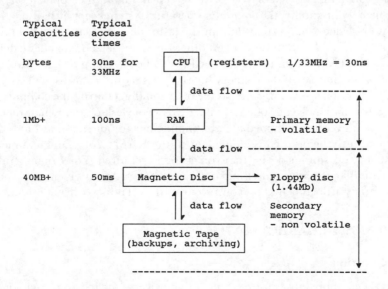

Figure 3.1 *Memory hierarchy*

Table 3.1 *Relative performance figures for various memory devices*

Memory device	Technology	Capacity/Mbytes	Access time	Cost per bit
CPU register	IC	0.00001	0.01 μs	$10
RAM	IC	1–10	0.1 μs	10 c
Disk	Rotating magnetic surface	10–100 +	0.01 s	0.01 c
Tape	Magnetic tape	10–100 +	1 s	0.001 c

with non-volatile magnetic disk and magnetic tape memory devices. Note that only data in the primary memory can be accessed by the CPU registers. Any data in the secondary memory must be brought into the primary memory before the CPU can access it. Moving down the hierarchy we see the following trends: decreasing cost per bit; increasing capacity; increasing access time; and decreasing frequency of data access to the CPU.

The relative performance figures for each type of device are shown in Table 3.1. Note that the continual improvement in the cost to performance ratio for magnetic media devices reflects the improvement in manufacturing techniques resulting from improved cost to performance ratios for ICs.

Random access memory is also called the *main memory*, *primary memory* or *immediate access store* (IS). Magnetic storage devices are called *secondary memory* or *backing store*. For maximum computer performance, the memory device one must be able to operate at a speed that is comparable to that of the CPU. In other words, a lot of primary memory is required. To operate at minimum cost must use magnetic storage devices, which unfortunately are very slow. There is therefore a trade off between data access time, cost, and capacity.

The relative mix of memory types varies depending on the application. For a given memory hierarchy, the computer operating system keeps little-used data in the slower devices and current data in the faster devices. Data stored on disk is brought into RAM only as needed. At other times it is held on magnetic devices. Anything in RAM that is no longer needed can be stored on disk for future use. Data flows up and down the memory hierarchy as required. As we will see in the next Chapter, the operating system is the program that controls the computer. An analogous situation is your office. The CPU is your desktop, on which there are papers (data) requiring immediate attention. Your in/out tray that holds documents currently being used is equivalent to the RAM. Your filing cabinet is a permanent bulk storage device that you access as required. You can only process documents that are on your desk and in the in/out tray. At any given time, you will have a variety of documents that you are currently attending to in your in/out tray.

3.2 MEMORY CHARACTERISTICS

3.2.1 Location

One of the most obvious features of memory devices is the location—where is the memory device situated? The main classification is *internal* to the computer box or

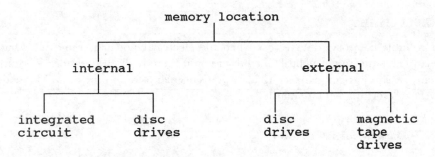

Figure 3.2 *Memory location*

external to the computer box. Typically, internal memory devices are ICs and disk drives. Because of physical restrictions in the size of the computer box, there can also be external memory in the form of tapes and disks. Disk drives may also be located external to the computer box, along with magnetic tape drives (Figure 3.2). External devices must access the computer through I/O controllers.

3.2.2 Memory Capacity

Memory capacity is expressed in bytes (1 byte = 8 bits). But, a word of warning: though less common now, memory capacity can also be expressed in words or even bits, where a word can be 8, 16, 32 or more bits.

3.2.3 Unit of Transfer

For the main memory, the unit of transfer is the number of bits transferred to or from the memory at a given time. Typically, the unit of transfer for the internal memory equals the number of data lines in the computer, which is usually equal to the computer's word length. For the example given in Chapter 3, the unit of transfer was 1 byte. The unit of transfer does not have to equal a word. For external memory, data is typically transferred in *data blocks* or blocks, units much larger than a word.

3.2.4 Performance

The access time is the time that elapses between the address being presented to the memory and the associated data being stored or read. Transfer rate is the rate at which data can be transferred into or out of the memory device.

3.2.5 Memory Technology

There are two common memory technologies. The first is the VLSI technology of RAM and ROM chips, which is used in semiconductor or IC memory. The second is the magnetic surface recording used by disk and tape drives.

3.2.6 Volatility

In volatile memory, data is lost when the electrical power is removed. Once recorded, data in non-volatile memory remain there without deterioration until deliberately changed. No electrical power is needed to retain the data. Magnetic surface recording is non-volatile.

3.2.7 Data-access Methods

The three methods for accessing data are random, direct, and sequential access.

In random access, each addressable location in the memory has a unique, physically wired-in addressing mechanism. The time to access a given location is therefore independent of any prior accesses. The time required for accessing and locating any data is a constant. Any location can be selected randomly, directly addressed and accessed. Integrated circuit RAM is random access.

In direct access individual data or record blocks have a unique address based on physical location. Direct access to the record or block is possible. Disk drives, which are direct access devices, are also referred to as cyclic access devices. The data in a cyclic access device circulates continuously in a repetitive loop and each data item is only accessible as it passes the reading and writing positions. The access time for cyclic access devices is therefore inherently longer since, on average, half a cycle must elapse before the desired data become available.

In sequential access, records can be grouped into memory blocks with no addresses. The access operation must be done in a specific linear sequence. Tape drives are sequential access devices.

3.3 INTEGRATED CIRCUIT MEMORY

Integrated circuit memory is classified as *random access*, having individual memory locations that are directly accessed through wired-in addressing logic.

Through historical accident, the most common type of semiconductor memory is also called RAM. This is a misuse of the term, as all types of IC memory are random access. The important feature of RAM is that data, in the form of binary electrical pulses, can be easily read from and written to the chip. Another characteristic is that it is volatile and as such should have a constant power supply. Even a temporary power loss will result in complete data loss. Random access memory is only used for temporary storage.

There are two principle RAM technologies, *static* RAM and *dynamic* RAM. A dynamic RAM (DRAM) stores data in individual binary cells. A binary cell in this case consists of a single capacitor and associated transistor. A charged capacitor represents binary 1; the absence of a charge represents binary 0. By nature, capacitors leak charge, consequently, dynamic RAM must be refreshed periodically to maintain the data conditions. This is done automatically. Static RAM (SRAM) employs cross-coupled transistors called flip-flops. The faster access time of SRAM is, however, offset by the lower storage density and significantly higher cost. Data is held for as long as power is supplied. Static and

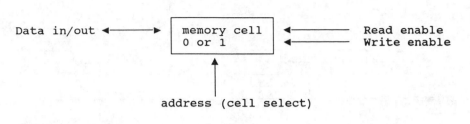

Figure 3.3 *The memory cell*

dynamic RAMs are both volatile. The dynamic RAM cell is a simpler construction and is therefore smaller. Smaller bit cells permit higher packing densities (higher levels of integration) and consequently show a decrease in cost. Static RAMs are, however, faster.

The ROM circuit that holds the permanent data/programs is called the *firmware*. Firmware can be read but not altered. A ROM device is an IC but the data/program is hardwired into the chip during fabrication. Hardwiring ensures that the data/program cannot be dynamically modified by the computer during operation. Read only memory chips are used to hold data and programs that must always be available to the computer, *e.g.*, part of the operating system.

Electrically programmable read only memory (EPROM) is reprogrammable and non-volatile. However, to be reprogrammed the device must first be removed from the computer or instrument that houses it and exposed to radiation for about 20 min. These devices therefore complement the ROM, which cannot be modified by the user. A further development is represented by the electrically erasable programmable read only memory (EEPROM), which can be reprogrammed whilst still in place. However, EEPROM is significantly more expensive. Do note that prices can change quickly.

The *memory cell* is the basic functional unit of semiconductor memory. The memory cell may be fabricated using different technologies, but will always have the following characteristics: two stable states that are used to represent binary 0 and binary 1 (equivalent to 1 bit); a unique address or location that allows selection of that particular cell; the state may be set (at least once—ROM, many times—RAM); and RAM cells can be selected for reading or writing.

This functionality is illustrated in Figure 3.3. The sequence of actions is as follows: the memory cell is selected using the address line; the read/write function is selected using the read/write select line; and the data are read from or written to the cell on the data line according to the preselected function.

Memory cells can be grouped together to form the functional unit called a *word*. In our example, the word is 8 bits long (b0 to b7). All the bits of one word have a common *word-select line*. Activation of the word-select line connects all the bits to the read/write circuits by the *bit lines*. When the word-select line is not activated the read/write circuits cannot access the word. After activation of the word select line, data from the data lines (D0 to D7) can be input to the word (*i.e.*, the write operation). Alternatively, data from the word can be output to the data lines (*i.e.*, the read operation) (Figure 3.4). Data lines are used for the

Word

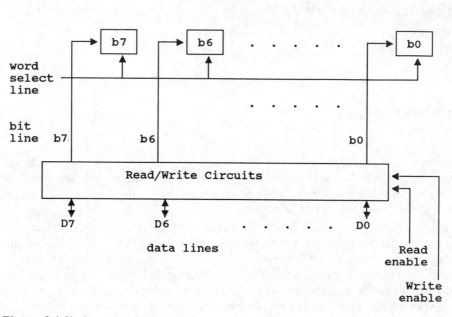

Figure 3.4 *Single word select circuit*

simultaneous input and output of data bits to and from a data buffer in the read/write circuits. On input (write-select function), the bit driver for each data line is activated to 0 or 1 according to the input data value. On output (read), the value of each bit line is put on the data lines. In both cases the word line selects which row (word) is used for reading or writing.

The number of words in the memory unit could be increased by having more word-select lines (Figure 3.5). In our case, there are 256 word-select lines (0 to 255 denary). Even though all the words have common bit lines, it is only when the appropriate word-select line is activated that data can be input or output to the word. Only one word-select line at a time is active. For example, activating word-select line 1 provides a connection between word 1 and the read/write circuits. All the other words are disconnected.

3.3.1 Random Access Memory

Let us consider what is sometimes called the 2D memory chip organization. In our 2D organization, the physical arrangement of the cells is the same as the logical arrangement of words in memory that we have seen previously. The data array is organized into W words of B bits each (Figure 3.6). We have (265×8) bit words (*i.e.*, a 2 Kbit chip or a 256 byte chip). In addition to our memory device, we need supporting circuits. The address lines supply the address of the word to be selected. A total of $\log_2 W$ lines are needed. In our example we have $\log_2 256$ or eight address lines (A0 to A7) to select one of 256 words. These lines are the

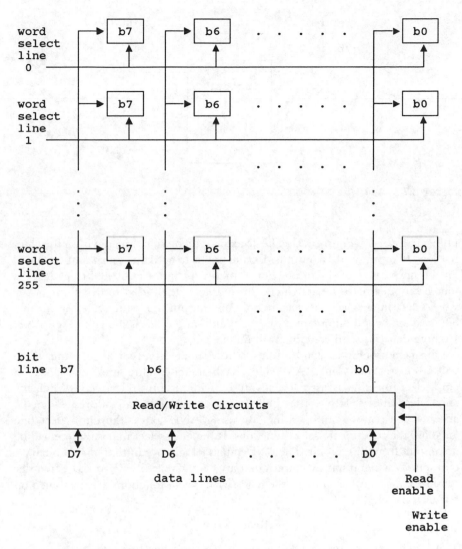

Figure 3.5 *Multiple word select circuit*

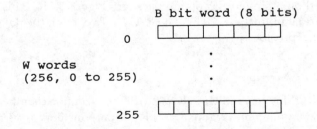

Figure 3.6 *W words of B bits each*

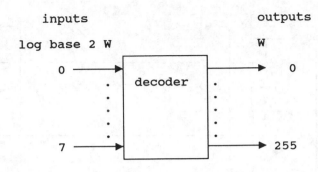

Figure 3.7 *Decoder*

inputs to a *decoder* (Figure 3.7). The decoder, which is contained within the chip, has $\log_2 W$ inputs and W outputs. Its function is to activate a *single* output based on the bit pattern of the input. For example, an input of 00000011 causes the fourth output line to be activated, this output is then used to drive one of the word-select lines. Decoders ensure only one output line is activated at a time. The resulting combination of the decoder and our circuit is shown in Figure 3.8. Do note that this is all contained within one chip.

Our memory chip is (256×8) bits, a little small. In a typical computer there will be more than one RAM chip. Each memory chip must then have a chip-select line that is used to activate individual chips selectively. When the chip-select line is active, that chip has been selected and addresses can be accessed for reading and writing. By using AND gates, the chip-select line 'enables' the read-/write-select functions. If the chip select line is 0 there will be no outputs from this circuit. The AND gate is a logic circuit that gives a binary 1 output only when it has two inputs that are each binary 1 (Figure 3.9). In order to reduce the number of lines, the read-/write-select functions are combined on one line and expressed as

$$\overline{\text{Read}}/\text{Write}$$

This statement says, read when line = 1 and write when line = 0.

The complete functional circuit for our (256×8) bit RAM, which is contained within one memory chip, is shown in Figure 3.10. The functionality discussed so far is illustrated on a pin-out RAM chip package (Figure 3.11). This configuration is called dual in-line pin (DIP). The RAM IC's contain the data memory map that we saw previously (Figure 3.12).

3.3.2 Read Only Memory

Functionally, ROM is very similar to RAM with the exception that the chip does not have the write function (Figure 3.13). Read only memory is used to store programs such as those that form part of the operating system. These will be considered in Chapter 4.

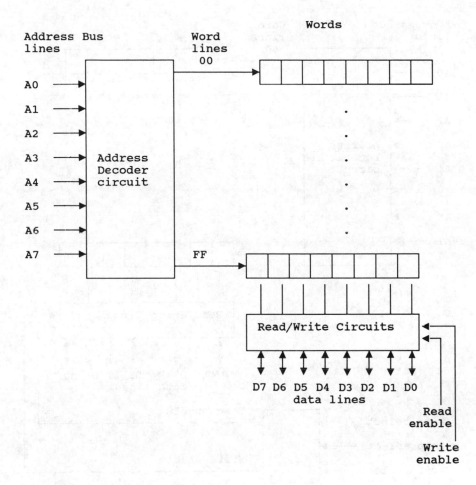

Figure 3.8 *Word select circuit with decoder*

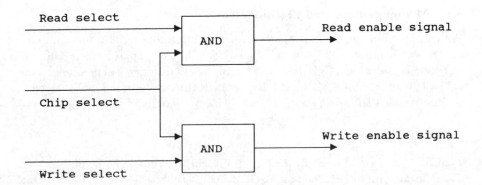

Figure 3.9 *Read/Write enable circuit*

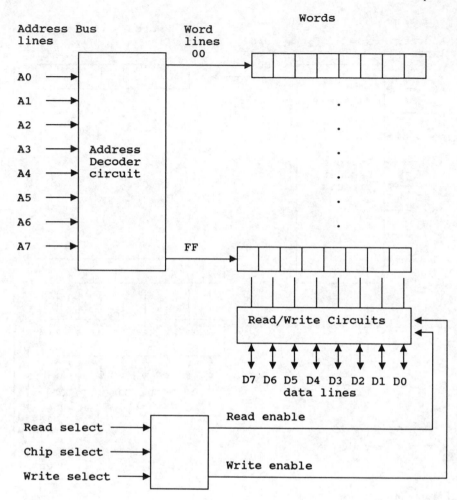

Figure 3.10 *Complete memory circuit*

3.3.3 Memory Integrated Circuit Selection

We can now implement a memory circuit for our computer. In our simplified model, we are using (1024×8) bit ICs with a 16 bit AB in place of our 8 bit AB. Selection between the ICs is made by the chip-select decoder—depending on the value of the address lines A10 to A15, any one of 64 ICs may be individually selected. Given that each IC is 1K, this gives us a memory capacity of 64K (Figure 3.14).

3.3.4 Memory Chips

Typically, memory chips are single, 1 bit chips. These are available with different capacities: $(16K \times 1)$ bit; $(64K \times 1)$ bit; $(128K \times 1)$ bit; $(256K \times 1)$ bit; $(512K \times 1)$ bit; $(1024K \times 1)$ bit.

The $(1024K \times 1)$ bit, which is called a 1 megabit chip, is typically used in single in-line memory modules (SIMMs).

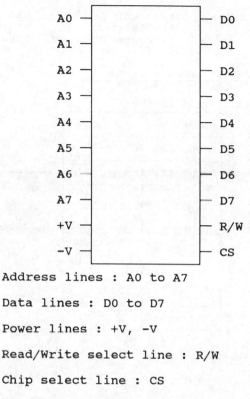

Address lines : A0 to A7

Data lines : D0 to D7

Power lines : +V, -V

Read/Write select line : R/W

Chip select line : CS

Figure 3.11 *RAM DIP chip*

Dual in-line pin memory chips are organized in banks, either on the mother-board or the memory board. The banks correspond to the DB capacity of the microprocessor. In our example, the banks would be 8 bits stored on 8 different chips, *i.e.*, one bit per chip. The IBM PC's version of ASCII has 8 bits (256 codes), the first half of which contains the standard ASCII character set. The remaining codes, from 128 to 255, consist of special characters and graphics. This is often referred to as extended ASCII. With parity, 9 bits are used and in this case we would have 9 chips. For 16 bit systems the banks are 14 bit plus 2 parity bits (*i.e.*, 16 chips wide). The 1 bit chip used to be a very common standard; however, it is now possible to see 2 bit or 4 bit chips.

As an alternative to individual chips many computers use SIMMs. A SIMM is a small p.c.b. with memory chips soldered onto it. This board plugs into a slot on either the motherboard or the memory board. As the memory chips are soldered onto the SIMM, they cannot be accessed individually. A single chip failure means that the entire SIMM must be replaced. Single in-line memory modules are very compact and are designed to eliminate 'chip creep'. This is the tendency for chips to work their way out of the sockets due to changes in temperature. This ultimately leads to poor contacts and memory failure. Soldering all the memory chips onto the board makes them difficult to replace. Servicing is therefore made more difficult, as the memory chips are often the most likely components to fail.

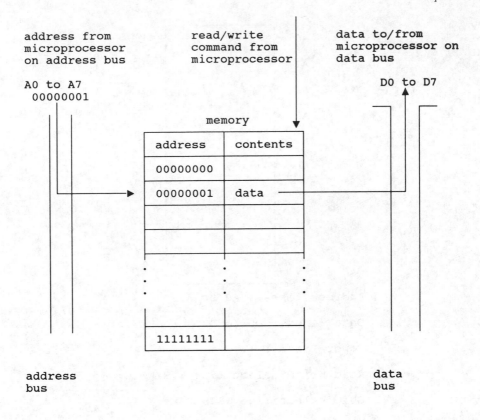

Figure 3.12 *Functional diagram of memory*

3.3.5 Cache Memory

The cache memory is an additional tier in the memory hierarchy (Figure 3.15). It is a very fast memory (SRAM typically, 20–30 ns), but restricted in capacity (64 Kb), and it is placed between the CPU and the RAM. The cache memory holds data that are often accessed by the CPU. The fast performance of the cache therefore considerably improves computer performance. Wait states refer to the mismatch between the CPU and memory speeds. For example, a CPU with a clock speed of 33 MHz has a clock period of about 30 ns (1/clock speed). A DRAM with access times of about 100 ns would cause the CPU to wait for three clock cycles, *i.e.*, three wait states. Using a SRAM with 30 ns access times results in zero wait states.

3.4 MAGNETIC SURFACE RECORDING

Magnetic tape and disk-storage devices depend on a constant relative motion between the read/write heads and a magnetizable surface. Information is stored by creating small magnetized areas of different polarity. In binary recording, the surface is divided into *bit cells* (Figure 3.16). A bit cell is the period during which

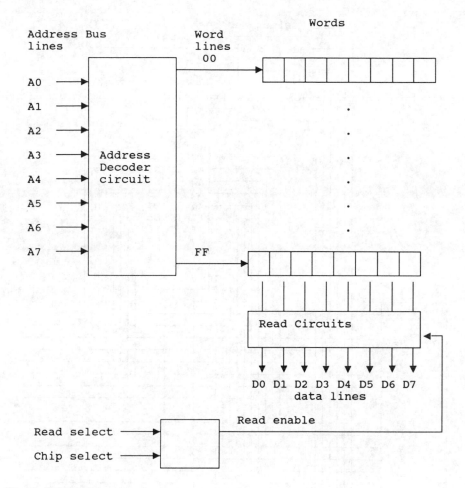

Figure 3.13 *ROM memory circuit*

a binary value is represented by two patterns of magnetization—one for binary 0, the other for binary 1. Recall that, in IC memory, the bit cell is either a capacitor or a transistor flip-flop. Data is transferred by means of read-write heads. A head consists of a high-permeability material around which a coil is wound, with a small air gap near the magnetic surface.

During the write operation, a current is passed through the coil, thereby generating a magnetic flux. The flux passes through the coil with leakage at the air gap and magnetizes the magnetizable surface beneath. Reversing the current reverses the direction of magnetization. The magnetized area can be considered as 1 bit of data. Moving the surface causes the process to be repeated (Figure 3.17).

In the read operation, the magnetic flux from the surface is diverted into the high-permeability head coil. A relative motion between the surface and the head will cause this flux to change. This will in turn induce an e.m.f. (voltage) across the coil that is proportional to the rate of change of flux (Faraday's law). The

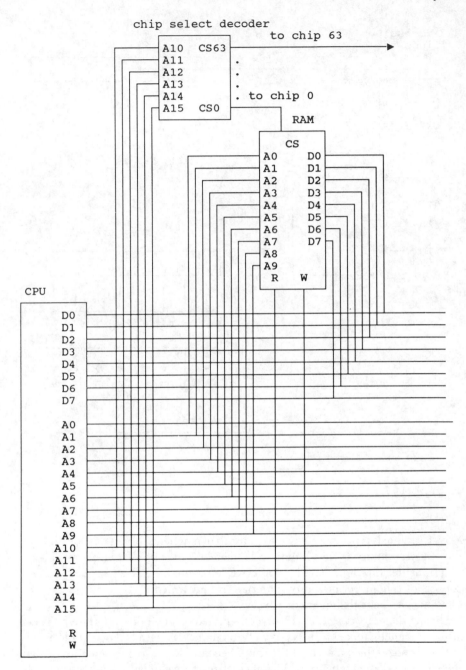

Figure 3.14 *Memory chip decoding circuit*

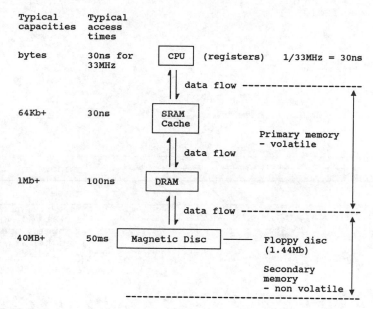

Typical capacities	Typical access times			
bytes	30ns for 33MHz	CPU	(registers)	1/33MHz = 30ns
		data flow		
64Kb+	30ns	SRAM Cache		
		data flow	Primary memory - volatile	
1Mb+	100ns	DRAM		
		data flow		
40MB+	50ms	Magnetic Disc	Floppy disc (1.44Mb)	
			Secondary memory - non volatile	

Figure 3.15 *Cache memory with zero 'wait states'*

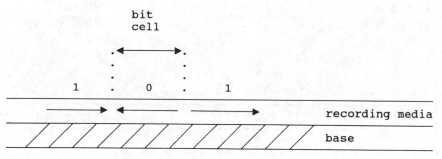

bit
cell

1 : 0 : 1

recording media

base

Figure 3.16 *The bit cell*

Applied field

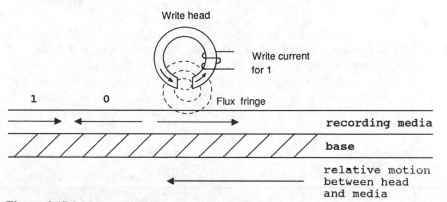

Write head

Write current
for 1

1   0

Flux fringe

recording media

base

relative motion
between head
and media

Figure 3.17 *Writing a bit cell*

Induced field

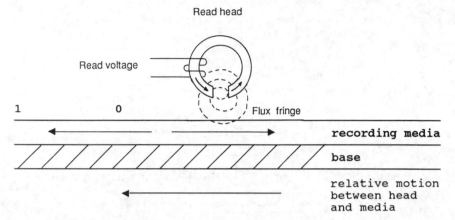

Figure 3.18 *Reading a bit cell*

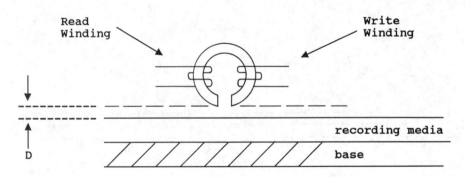

Figure 3.19 *A combined read/write head*

polarity of the induced voltage depends on the polarity of the magnetic field (Figure 3.18).

In practice, the read/write functions are performed on one read/write head (Figure 3.19). The bit cells are sequential on unconnected sequential *tracks*. The bit cells should be as small as possible in order to achieve a high packing density. This economizes on the surface used and minimizes the time taken to transfer data. A primary consideration is how to minimize the gap shown as D in Figure 3.19. This is achieved in different ways depending on the device that is being used.

3.4.1 Recording Techniques

The requirement for having maximum storage capacity and data transfer rate, is to have as many bit cells per unit length as possible. This must not be at the

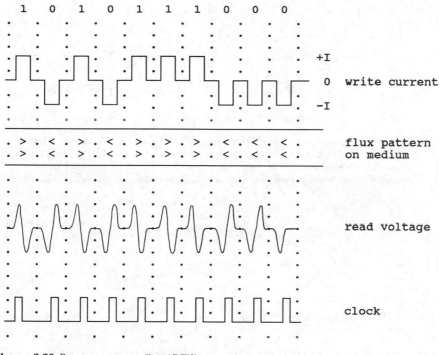

Figure 3.20 *Return to zero encoding (RTZ)*

expense of reliable data transfer. It is therefore necessary to make bit cells as small as possible and to minimize overlap between bit cells. The data transfer must also be synchronized. If a long sequence of 1s produces a constant output signal, how can we accurately determine where each bit boundary is? A guaranteed regular *clock pulse* is needed to define the bit-cell period. In some cases, a separate clocking track is used—*multi-tracking*. However, some magnetic storage devices (*e.g.*, some types of magnetic disk) are designed to access only one track at a time. This track must therefore contain both the data and the clock signals. Recording techniques that contain at least one flux reversal per bit cell do not need this separate clocking track and are classified as *self-clocking* methods. Recording methods that ensure that there are no unmagnetized areas, employ what is called *saturation* recording. With *unsaturated* recording methods, the magnetic field returns to zero when a digit 0 or 1 is not present. In this case, it is very difficult to overwrite data that has been written previously unless the position of each cell is accurately located—a laborious task. Alternatively, all flux can be erased before writing new data, but this requires an additional erase head. Two data recording methods will now be considered.

3.4.1.1 Return to Zero. In the return to zero (RTZ) method, current passes through the write winding only when a digit 0 or 1 is to be recorded. If a digit 1 is to be recorded, a current pulse of positive polarity is applied. Conversely, if a digit 0 is to be recorded, a negative pulse is applied. In both cases, the current in

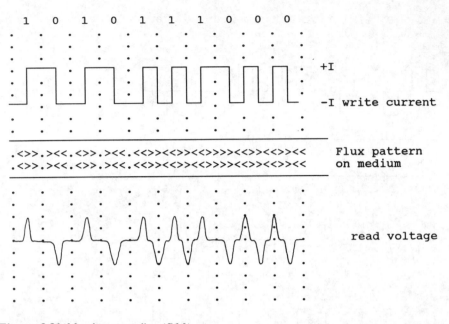

Figure 3.21 *Manchester encoding (PM)*

the write winding is *returned to zero* after the pulse passes, and remains so until the next pulse is applied. Hence, small magnetic cells are induced in the magnetic surface, the flux direction representing the two digit values. If this pattern is passed under a read head, a pulse is induced in the read winding every time a change of magnetic flux is encountered (Figure 3.20). A clock pulse is needed to determine the exact position of the bit cell.

3.4.1.2 Phase Modulation—Manchester Encoding. In the phase modulation (PM) method, half of each bit cell is magnetized in opposite directions. A digit 0, for example, is recorded as a half-bit time-positive pulse followed by a half-bit time-negative pulse. A digit 1 is recorded as a half-bit time-negative pulse followed by a half-bit time-positive pulse. The data are thus represented by the direction of pulse transition in mid-bit time, where a digit 1 has a positive leading edge and digit 0, a negative edge (Figure 3.21). The easiest way to remember this method is to memorize that 1s have a low–high transition in the middle of the bit cell, and 0s have a high–low transition in the middle of the bit cell. To allow for continuous streams of the same type of digit there are bit cell boundary transitions. This method is self clocking.

3.5 MAGNETIC DISK STORAGE

Magnetic disks are undoubtedly the most popular form of on-line storage for use with both large and small computers. Magnetic disk drivers are available in various forms, from the high capacity, multi-platter hard-disk drives commonly found in mainframe and minicomputer installations, to the tiny 3.5″ drives used in the smaller PCs. A magnetic disk is a circular platter made of either metal or plastic that is coated with magnetizable material. Data are recorded onto and can be retrieved from a series of concentric tracks on the disk by means of a read/write head. During read/write operations, the read/write head is stationary while the disk rotates beneath it—there is relative motion between the disk and the read/write head.

3.5.1 Rigid Magnetic Disks

A typical hard disk has one or more platters or disks, for which the two most important diameter sizes are 5.25" and 3.5" (Figure 3.22). Larger drives with 8" or 14" platters are available on minicomputers. Recent trends have been towards platters smaller than 3.5". Each platter is made of aluminium, typically 0.125" thick, and coated with a magnetically responsive surface. Hard disks are usually fixed and cannot be removed from the computer. Removable disk packs are used on large computer systems. Three sizes of drive are available for the PC: 5.25", full height; 5.25", half height; and 3.5", half height.

3.5.1.1 Flying Read/Write Heads. A hard disk usually has one read/write head for each side of each platter. The multiple heads are connected or ganged together on a single movement mechanism, the comb (Figure 3.22). All the heads move in unison across the platters. Each head is on an arm that is spring loaded to force the heads onto the platter surface. Each platter is squeezed by the two heads above and below it. Disks rotate at very high speed, typically 3600 r.p.m. Recall that we wish to make the bit cell as small as possible. We can do this by making the head as small as possible and applying only small currents to the heads. However, for reliable data transfer, this means that the head must be as close to the surface as possible without actually touching it. A head in contact with a surface rotating at 3600 r.p.m. would not last very long! A typical distance between the head and the disc surface is 2 μm or less. Further, the head must be maintained at this constant distance above the surface of the disk. If the gap were to increase temporarily, data transfer would be unreliable. If the head were to come too close to the surface this could result in destructive contact between the two—'*a head crash*'. The distance of the head from the surface of the platter is therefore critical; it must be very small and constant. At these values the disk surface will not be smooth, it will undulate. It is not economically possible to engineer such a system directly.

How is it done then? When the disk rotates at its operational high speed, a boundary layer of air builds up on its surface. The aerodynamically shaped head rides on this cushion of air, hence the term a *flying head* (Figure 3.23). The head follows the undulations of the disk surface. This prevents wear on the disk and head surfaces yet maintains a constant magnetic coupling between the head and disk. The dimensions are put in perspective when one considers the damage that hair and smoke particles can do to a disk (Figure 3.24). To exclude particles of dust, dirt and smoke (which may otherwise lodge between the head and the disk surface), the entire disk assembly is housed in a tightly sealed enclosure. At rest,

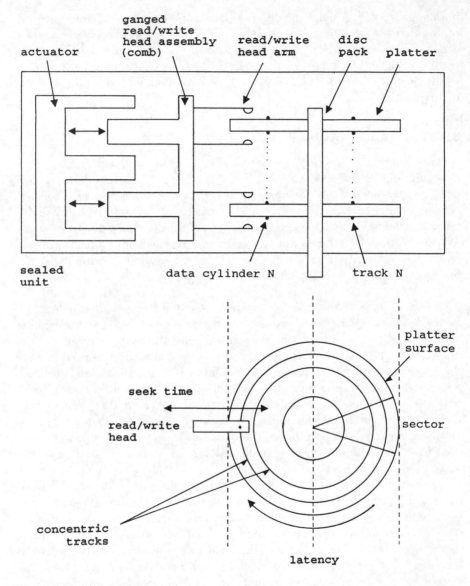

Figure 3.22 *Hard disc drive components*

the heads sit on the platter. Due to the action of the springs forcing the head towards the disk surface and the counterbalancing aerodynamic force, disk drives can be operated, not only on their side, but also upside down.

Unfortunately, disk drives need careful handling as permanent damage can result if the drive is bumped or jolted (either when in use or in transit). This is likely to cause the head to 'crash' against the oxide surface of the disk, which not only may damage the disk surface, but could also ruin the drive's read/write head or at least misalign it so as to prevent correct reading and writing. Some

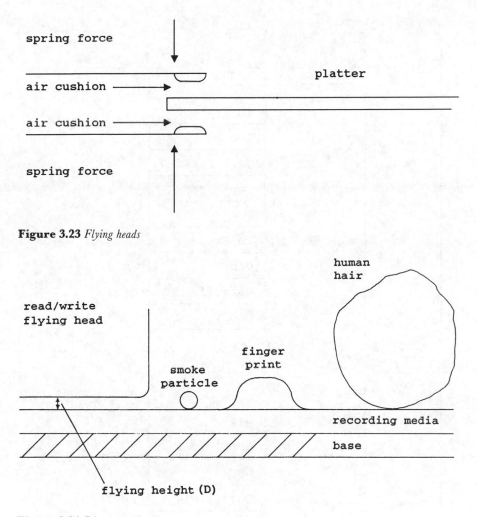

Figure 3.23 *Flying heads*

Figure 3.24 *Disc contamination*

manufacturers allow what is called 'parking' of the heads through a software command, others have a system whereby the head is automatically parked whenever the drive is not in use.

Always buy a hard disk that is *shock mounted*. This means that there is a rubber cushion between the disk drive body and the mounting chassis. This has the effect of cushioning the heads from any sudden shocks to the computer.

3.5.1.2 Data-access Time. Any particular address on a track becomes available once per revolution—*cyclic access*. A disk is often called a *direct-access device*, as it is not necessary for all the sectors on the disk to be read sequentially to reach the desired one. It must only wait for the intervening sectors within one track. In most disk systems, there is only one read/write head per disk surface and therefore the head assembly has to be positioned by a servomechanism over the

Index   Sector

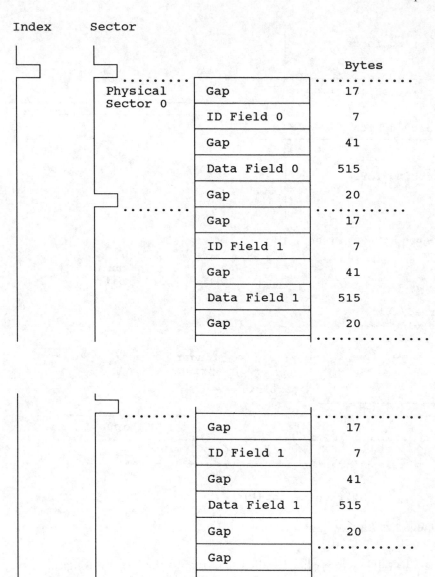

Figure 3.25 *Track formatting*

required track. The tracks are arranged as concentric rings on the surface of each platter. The time taken to access the desired track is called the *seek time* (Figure 3.22). Time is also required for the disk to rotate to the point under the head assembly where the data are. This is called the *latency time* (Figure 3.22). The access time is therefore the sum of the seek and latency times, typical values of which are 30 and 10 ms, respectively. The *transfer rate* is the speed at which the data can be transferred to/from the disk pack.

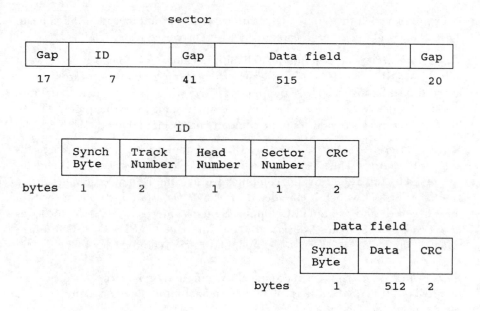

Figure 3.26 *Sector information*

3.5.1.3 Data Cylinders. As we have seen, the head assembly consists of a comb-like structure that moves in and out to select the required track, but on different surfaces. All the heads are synchronized so that at any instance they are all aligned to the same track, but on different surfaces. This set of tracks is called a *cylinder* of information (Figure 3.22). Data are normally written to, or read from, each track of a cylinder in turn before the head assembly is repositioned to another cylinder. By selecting cylinders of information in this way, mechanical movement of the heads (the seek time) is minimized.

3.5.1.4 Disk-track Format. Rather than addressing each byte or word, each track is divided into individually addressed sectors that typically hold 512 bytes of data. There are usually between 8 and 100 sectors per track. Unformatted disks do not have these addressable sectors, so disks must be formatted before they can be used. To identify each sector, the associated control data are stored on the disk (Figure 3.25, Figure 3.26). It is therefore necessary to format the disk with this extra data that used only by the disk driver and are not accessed by the user. Owing to the geometry of the concentric recording tracks, the track length reduces as the track radius decreases. The packing density is therefore higher on the inner tracks.

Data are transferred to and from the disk in blocks. On the disk, data are stored in block-sized regions called sectors. For example, each track contains 30 fixed-length sectors of 600 bytes each. Each sector holds 512 bytes of data plus the control information used by the disk controller. The ID field is a unique identifier (address) used to locate a particular sector on a track. The SYNCH byte is a special bit pattern that delimits the beginning of the data field. The track number identifies a track on a surface, the head number identifies a head. The ID and data fields each contain an error correcting code (CRC).

3.5.1.5 Reliability. The criterion for reliability evaluation is mean time between failure (MTBF). Valves of MTBF range from about 20 000 h upwards. These are statistical figures. They are put in perspective when one considers that a disk working 24 h/day will operate for 8760 h/year. In some cases the quoted MTBF figures are in excess of how long the company has been making disks!

3.5.1.6 Interface Specifications. There are various popular interfaces that are available for PC hard-disk drives: the ST, which was developed by Seagate Technologies in 1980 (this is 'old' in the computer world); the Enhanced Small Device Interface (ESDI), which is considered to be a *de facto* standard, but not within the PC world; and the Small Computer Systems Interface (SCSI), which was preceded by the Shugart Associates System Interface (SASI). The SCSI, affectionately known as 'scuzzy', is a very popular *de facto* standard for PCs.

3.5.1.7 Head Actuators. The mechanical system that moves the heads to the required position on the disk is called the head actuator. The two main types are the stepper motor and the voice coil.

A stepper motor is an electrical motor that can move through a series of pre-set positions or steps. If you ever have the opportunity to hold one and move the comb, you should be able to feel the stepping effect. The closest analogy is to be found in stereo systems that have a volume control with a stepping effect. Each step represents a track position. The biggest problem with stepper motors is temperature. The metal platters or disks expand and contract with temperature. This causes the tracks to move in relation to an externally defined track position. The stepper motor cannot move in the increments of less than a single track that are necessary to compensate for track movement.

High-quality, high-performance hard disks use voice coil actuators, which are similar to the construction found in loud speakers. In hard-disk systems, the electromagnetic coil moves on a track through a stationary magnet. There is no physical contact between the coil and the magnet, only magnetic coupling. The coil mechanism is connected to the head comb. Energizing the coils results in either attracting or repulsing the magnet, thereby causing the head comb to move. This type of device is quicker than a stepper motor. It does not have predefined step positions, therefore, additional features are needed. During manfuacture one of the sides of one of the platters is not used for reading/writing but reserved to hold control data. On it are marks that indicate the correct track positions. Associated with this surface is a read-only head. This read head reads the control data from this surface to position the heads. This is called the dedicated-surface, closed-loop, servo-controlled mechanism. Dedicated surface refers to the platter surface reserved for control data. It is for this reason that voice coil drives have an odd number of heads. This type of mechanism is not affected by temperature changes. The voice coil automatically compensates for any changes in the track positions. Voice coils provide automatic head parking.

Actuator type has a direct bearing on drive performance. Things do change, and especially in the computer field. Typically, cheaper, low-capacity hard disks use a stepper motor, and high-performance and high-capacity hard disks use a voice coil, *i.e.*, a low mass moving coil actuator with embedded servosystem.

3.5.1.8 Head Parking. When a hard disk stops, the heads land on the disk. Usually they land wherever they happen to be when the power is removed, even on data-storage tracks.

Drives with a voice actuator coil provide automatic head parking so that when the power is cut the heads are safely positioned or removed from the disk.

Typically, drives with stepper motors do not park heads automatically and parking is accomplished by the software. The software is a small program that secures the heads. Unfortunately, power must be supplied for the software to run, a catastrophic power failure will mean that the heads will not have been parked.

3.5.1.9 Air Filters. A hard disk is a permanently sealed unit. The recirculating filter contained within the unit is designed to filter off small particles of media that have been scraped off the surface during head landing and take off. Being sealed units, disk drives are able to function in harsh, dirty environments. All drives have a pressure equalization system with a filter port that bleeds air into and out of the drive to compensate for large changes in atmospheric pressure. The filter used has a high-quality submicron specification. It is interesting to note that the term 'Winchester' was often associated with hard disks. This usage has its origins in the history of computing. In the 1960s, IBM developed a combined 30M removable- and 30 Mb fixed-disk system with flying heads. This drive became known as the 30–30 drive, and, very quickly, the 'Winchester' after the Winchester 30–30 rifle.

3.5.1.10 Prairie Tek 2.5″ Hard-disk Drive. We should now be able to examine a technical specification critically. The following are extracts from a technical specification.

Dimensions	2.8″ wide, 4.3″ long, 1″ high
Capacity	per drive, 20 Mbytes (formatted)
	per track, 8192 bytes
	per sector, 512 bytes
	sectors per track, 16
Interface	SCSI
Configuration	Number of disks, 2
	Number of data surfaces, 2
	Number of data heads, 2
	Number of data cylinders, 612
Performance	Rotational speed, 3367 r.p.m.
	Data transfer rate, 0.625 Mbytes s^{-1}
	Rotational latency, 8.9 ms
	Average seek time, 28 ms
	Track-to-track seek time, 10 ms
	Maximum seek time, 45 ms

3.5.2 Floppies

The 'floppy' disk is made of a soft, flexible material called Mylar, which is coated with magnetizable material. Mylar is a dimensionally stable plastic. A floppy-disk drive, irrespective of its type, consists of several basic components.

3.5.2.1 Read/Write Heads. When the world was young, single-sided drives were available for the PC. Almost any disk drive purchased today will be a double-sided drive with two read/write heads. Each side of the disk has its own head and both heads are used for reading and writing. The two heads are mounted on the same rack mechanism and are therefore moved in and out together. Movement is controlled by means of a *head actuator*. The heads cannot move independently of one another. The heads are made of soft ferrous compounds surrounded by electromagnetic coils. In fact each head is a composite construction with a read-write head contained within two erase heads. This type of construction allows the *tunnel erasure* recording method (Figure 3.27). As each track of data is written onto the surface, the outer erase heads erase the outer bands of the track thereby removing any bit cells. This has the effect of preventing magnetic signals from other tracks from affecting each other. The heads in effect force the data to sit in a narrow tunnel. Alignment is the placement of the heads in the correct position to read/write data. Head alignment can only be checked against a standard reference disk, which is available. The two heads are spring loaded with the effect that the disk is lightly gripped between them. Floppy disks rotate from about 300 to 360 r.p.m., which prevents frictional problems. Disks are typically coated with low-friction compounds to further reduce friction. The head/disk contact does, however, lead to a build up of surface material on the heads over a period of time. This accumulation of oxide should be removed periodically using a cleaning disk and fluid, for which kits are available.

3.5.2.2 Head Actuator. The head actuator is a mechanical motor that causes the heads to move in and out over the disk surface. Almost all floppy-disk drives use *stepper motors*, which move in fixed increments as opposed to being infinitely variable. Each increment therefore defines a track position. For example, to position the head at track 15, the motor is incremented 15 times. The speed of the stepper motor determines the track access time.

3.5.2.3 Spindle Motor. The spindle motor is used to rotate the disk, typically at 300 or 360 r.p.m. Earlier drives were coupled to the disk spindle by means of a belt. Most modern drives use a more reliable, cheaper, and smaller direct-drive system.

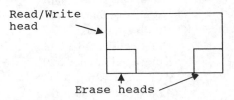

Figure 3.27 *Floppy disc read/write head*

Table 3.2 *Track widths*

Drive type	Number of tracks per side	Track width/ mm
5.25″, 360K	40	0.330
5.25″, 1.2M	80	0.160
3.5″, 720K	80	0.115
3.5″, 1.44M	80	0.115

3.5.2.4 Physical Operation. The floppy disk is rotated at either 300 or 360 r.p.m. While the disk is spinning, the read-write heads are moved in and out to access the different tracks. The heads have a total travel of about 1″ to cover either 40 or 80 tracks. The tracks are on both sides of the disk and are called cylinders. A single cylinder comprises the corresponding tracks on both surfaces. The tracks are recorded at different widths, depending on the type of drive (Table 3.2).

3.5.2.5 Logical Operation. From the perspective of the disk operating system (DOS), the data on disks are logically organized into tracks and sectors. Disk formatting depends on the DOS version used. Disk operating systems 3.3 and above support the 3.5″ disks having 18 sectors and 80 tracks (Table 3.3). The track nearest the outside edge of a disk (track 0) is reserved by the DOS for its own purpose. Track 0 in sector 1 contains the 'boot record' or boot sector that the system needs to begin operation. The next few sectors contain the file allocation table determining which portions of the disk contain information and which portions are empty. The next few sectors contain the directory information about the files. This is all 'user transparent'.

3.5.2.6 Clusters. The DOS stores files on disks in groups of sectors called *clusters* or *allocation units*. A cluster is the smallest unit that DOS can deal with when reading/writing to disks. A cluster is one or more sectors. Having more than one sector per cluster reduces the size of the file allocation table (FAT) and thereby improves performance. The disadvantage is the additional disk space that is wasted. Depending on its size, the file may fit into a single cluster or require more than one. The DOS attempts to store all the clusters associated with a file in

Table 3.3 *Floppy-disk formats (DOS 3.3 and above)*

Disk width/ inches	Density	Number of bytes per sector	Number of sectors per track	Number of tracks per side	Number of sides	Capacity/ K	Track width/ mm
5.25	Double	512	9	40	2	360	0.330
	High	512	15	80	2	1200	0.160
3.5	Double	512	9	80	2	720	0.115
	High	512	18	80	2	1440	0.115

Table 3.4 *Clusters*

Drive	Default cluster size	Number of clusters
5.25″, 360K	2 sectors (1024 bytes)	354
5.25″, 1.2M	1 sector (512 bytes)	2371
3.5″, 720K	2 sectors (1024 bytes)	713
3.5″, 1.44M	1 sector (512 bytes)	2847

consecutive disk locations. When this is not possible fragmentation occurs leading to unused capacity (Table 3.4).

3.5.2.7 Coercivity. The coercivity specification of a disk refers to the magnetic field strength required to make a reliable recording. Coercivity is the magnetic field strength measured in oersteds. A disk with a higher coercivity rating requires a stronger magnetic field to write data onto the disk. A disk with a lower coercivity rating can be written to with a smaller magnetic field. Lower coercivity means greater sensitivity. With increasing demands for bulk data storage with higher bit densities, high-density drives were introduced. However, the densities of the magnetic bit cells became such that adjacent bit cells began to affect each other to the extent that data erased itself over time. A simple analogy is provided by a line of magnetized ball bearings. If the spacing between the bearings is adequate, the individual magnetic fields will be too weak to interfere with each other. However, as the distance between the bearings is decreased, the magnetic fields will start to intefere to the extent that adjacent bearings will begin to be attracted to each other and to move. This migration and clumping can be eliminated by reducing the magnetic field of each ball. The reduced magnetic field allows the balls to be placed closer together.

Attempts to format a high-density disk with a double-density format may fail. The double-density fields are too low to affect the magnetic surface of the high-density disk. However, it is possible to format a double-density disk as if it were a high-density disk. There may be a higher proportion of bad sectors, but formatting should still be possible. High-density disks are also called *high-coercivity* disks as they require a much higher magnetic field strength than double density disks. The 360K disks require a magnetic field strength of about 300 oersteds, whereas high-density discs require a magnetic field strength of about 600 oersteds. Originally, the 5.25″ disks used FM recording; the introduction of MFM and MMFM led to double-density capacity.

Density is the amount of data that can be reliably stored on a given area of recording surface (Table 3.5). Two types of density exist, longitudinal and linear density. Longitudinal density, which is a measure of how many tracks there are per surface, is measured in tracks per inch (TPI). Linear density is a measure of the number of bits per inch (BPI).

3.5.2.8 Types of Floppy-disk Drive. The 8″ floppy-disk drive was originally developed by IBM as part of the IBM key-to-disk data-entry system. These were the forerunners of the 5.25″ mini floppy and had storage capacities of between 250

Table 3.5 *High-density and double-density disks*

Disc width/ inches	Media type	Number of tracks per inch TPI	Number of bits per inch BPI	Coercivity/ oersteds
5.25	Double density	48	5876	300
	High density	96	9646	600
3.5	Double density	135	8717	300
	High density	135	17434	600

Kbytes and 2 Mbytes. In recent years, however, the 8″ disk has rapidly faded from popularity, having been replaced by lower-cost, smaller floppy disks.

Mini floppy-disk drives are often categorized by their height in relation to those that were first available (*i.e.*, full-height drives). In recent years, there has been a shift towards more compact drives that take up less space in the equipment into which they are fitted. This is particularly important where drives are to be fitted into portable equipment. Consequently, half-height drives have also become increasingly popular. The 5.25″ floppy disk has a *central hub* that is grasped by the drive mechanism. This point on the disk is usually reinforced to extend the mechanical life of the disk. The smaller *index hole* gives the drive access to sectoring information. Careful manual rotation of the disk within the sleeve will reveal a small hole on the disk itself. This hole is used by the drive as the starting point for all disk sectors. If there is only one hole, the disk is soft sectored with the software determining the number of sectors. More unusually, the disk may have many such holes indicating that it is hard sectored; these are not usually used on PCs. The *media access hole* allows access to the disk surface. It is through this hole that the read/write heads contact the disk. The read/write hole is a notch in the side of the cover. If the notch is present, the disk is write enabled; when the notch is covered, the disk is write protected.

The 3.5″ disk system is based on a disk the surface area of which is less than half that of its 5.25″ counterpart. These disks are housed in a rigid plastic case, which helps to stabilise the disk, and hence improve performance. The disk is provided with an automatic shutter that protects the magnetic surface from accidental contact (a perennial problem with the 5.25″ mini floppy disk). The metal shutter, which protects the media access hole, is opened by the drive on insertion. There is a metal centre hub with an alignment hole. The disk is held within the drive by the hub and the alignment hole. On the plastic case there is a write *protect/enable hole*. With the slider positioned so that the hole is visible, the drive cannot write onto the disk (*i.e.*, the disk is write protected). When the slider is positioned so that the hole is not visible the drive can write onto the disk (*i.e.*, the disk is write enabled). On the other side of the jacket to the protect/enable hole is the *media density selector hole*. If this hole is present, the disk is a high-density (600 oersted) disk. In the absence of this hole, the disk is a low-density (300 oersted) disk. Many 3.5″ drives have a media sensor that controls the recording depending on the media density selector hole.

Both 3.5″ and 5.25″ disks are constructed of the same material—Mylar coated with a magnetic compound. The type of magnetic compound used determines the type of disk. Double-density disks are fabricated from an iron oxide based compound, high-density disks are based on cobalt.

3.5.2.9 General Information—Care, Repair and Use. Consideration of how disks are made and how they operate helps to explain the following rules of use: never touch the surface of disks; never write on disks with a pen or pencil; never bend disks; never allow disks to get hot; and never expose disks to strong magnetic fields.

Contrary to many beliefs, *X*-ray machines do not damage computers. However, as metal detectors produce very strong magnetic fields that could well affect magnetic media, it is always best to check at airports.

A point that is often forgotten is that the read/write heads make physical contact with the surface of floppy disks. As the disk is rotated at about 360 r.p.m. floppy disks do wear out. As a general rule it is advisable to replace on a regular basis floppies that are used frequently.

3.6 MAGNETIC TAPE

The way in which a storage device operates affects the speed with which any given data item held on the device can be located. Magnetic-tape drives are serial devices in that all the data items preceding the required one have to be traversed serially to locate the desired item. If a tape happens to be positioned so that we currently have access to the first data item on the tape and the required record is the last one, it will take several minutes to reach it. To access the desired record, all intervening records will have to be scanned by the reading heads even though they are not required. This is called *sequential processing*. Note that unlike disks, the tape is only in motion during read/write operations.

Magnetic-tape units have been developed and improved over many years so that they now provide one of the most robust and reliable types of backing storage device available. Magnetic tapes are relatively cheap, hold large data volumes and are ideally suited to off-line storage. Magnetic tape is also an important input medium because data on other media are often placed on magnetic tape prior to input to other large computers. Industrial standards apply to 'Mag' tape recording so that it is relatively easy to transfer large volumes of data from one computer to another. These industrial standards specify how records or block of records are written to tape. The standards are defined by, amongst others, the American National Standards Institute (ANSI). The standards define factors such as tape dimensions, tape speed, bit density, recording code, error-detection methods, and block size (minimum and maximum).

Magnetic-tape devices, though less common now, still provide a universal means of cheap, mass storage. It is anticipated that they will continue to have an importance in electronic data processing. Main uses include the following: archiving of data for long-term storage; back-up of data for security; and data capture by means of incremental data logging.

Typically, tapes are 2400′ to 3600′ long. Data are stored on parallel tracks on one side of the tape only. The number of tracks depends on the tape width. These are read/written to as the tape passes a head assembly. The tape is available in various sizes between 0.38 and 2.54 cm, and is about 0.025 mm thick. The tape is wound on reels that are, typically, up to about 30 cm in diameter. To detect the end of the tape, there are physical markers at the end in the form of metallic foil or a series of holes that can be sensed. Typically, NRZI recording is used.

3.6.1 Reel-to-reel Tape Transport

Unlike the disk, the tape is only in motion during the read/write operation. Data cannot be read or written until the tape is moving at the correct speed. To

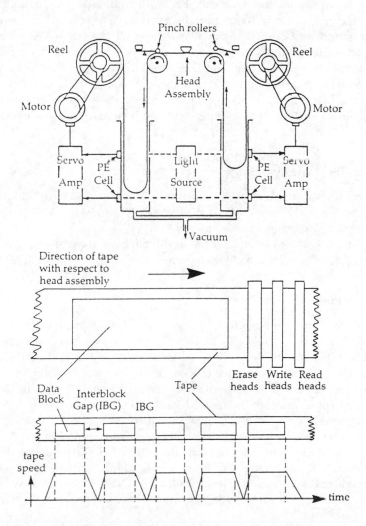

Figure 3.28 *Reel-to-reel tape mechanism diagram*

minimize wastage it is essential to accelerate the tape to speed as fast as possible and to decelerate as fast as possible. Data are organized and transferred to or from the tape in blocks under program control. To ensure that the start and stop times are not limited by the acceleration time of large, heavy, fully wound reels of tape, reservoirs of tape are provided to allow high acceleration at the head assembly. The component parts of a typical tape-transport mechanism are illustrated in Figure 3.28. The tape passes over two capstans rotating in opposite directions at constant speed, and is held in contact with the head assembly, which is usually mounted between the capstans. If the tape is being transferred from the right to the left spool, the pinch roller is brought into contact with the tape as it passes over the left capstan, causing the tape to move from right to left. When enough tape has been drawn from the reservoir on the right to allow activation of the top photoelectric cell (PEC), the motor driving the right-hand reel switches on, and unwinds tape into the reservoir. The reservoir fills with tape until the bottom PEC is de-activated, switching off the motor. When enough tape has been fed into the reservoir on the left to de-activate the lower PEC, the motor operates to wind the tape onto the left-hand reel. By applying a partial vacuum to the bottom of the reservoirs, the tape loops are forced down, thus creating spooling tension. Only the pinch rollers come into contact with the magnetic oxide surface. These rollers can be eliminated by using vacuum capstans.

3.6.2 Smaller Tape Systems

In smaller tape systems, the reels of tape are only a few centimetres in diameter making vacuum chambers redundant. The two main methods of tape transportation are by capstans, and reel-to-reel.

The reels are enclosed in an easy-to-handle package. Special digital cassettes (cartridges) have been designed for computer applications (Figure 3.29).

3.6.3 Data Format

Tracks of data are stored along the length of the tape and there is one read/write head per track position. The blocks can be identified either by a separate block-marker track, or by an address word preceding (or terminating) the data block. Interblock gaps (IBGs) allow the acceleration of tape to operating speeds before data transfer (Figures 3.28 and 3.30). No data are stored in the IBGs. Imagine listening to an audio tape at the wrong speed! For synchronization purposes, a clock track can also be included, but self-clocking is also employed. Data checking is normally performed by either parity or sum check or both. Parity is usually checked character by character across the track using a separate track for the parity bit. A longitudinal parity check character is also produced at the end of each block. A CRC character may also be generated at the end of each block. This is the summation of all the bits in the block (Figure 3.30).

Often, information on tape may not be read for a considerable time, hence it is

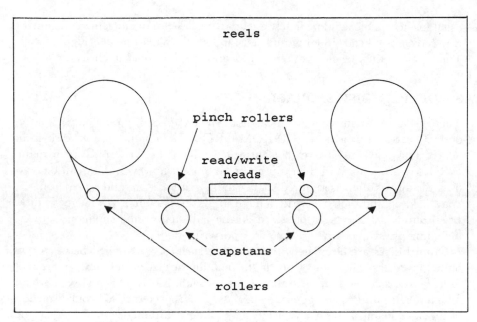

Figure 3.29 *Cassette diagram*

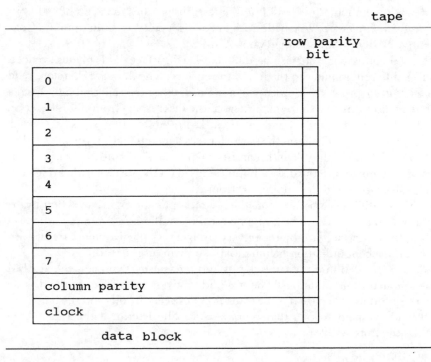

Figure 3.30 *Tape error detection*

usual to check a block immediately after writing. The most efficient arrangement is to have erase heads, then write heads, followed by read heads (Figure 3.30). The tape can then be cleared, written to, and the data read to check.

3.7 OPTICAL-DISK STORAGE

There are three main types of optical disk. The compact disk read only memory, optical read only memory (CD-ROM/OROM) is a non-erasable disk from which data may only be read. Data is written to the write once read many (WORM) disk once and then can never be erased; however, data can be read from the disk as often as required. Data can be written, read, and erased from the erasable disk, and then overwritten if required; this is similar to a magnetic disk. Irrespective of the type of optical disk, they all share some common features, including physical characteristics and read/write modes.

Optical disks are circular, with a metallic surface sandwiched between two outer glass disks. The composition of the metallic surface varies between types of disk. Plastic may be used instead of glass, providing it is of a quality that can match the fine optical characteristics of glass. This is necessary to avoid distortion of the laser beam. Disks are typically 12″ diameter although other sizes can be found, for example: 14″, 8″, 5.25″, and 3.5″. Diode lasers are used for data transfer. The laser focuses under the surface and is therefore less sensitive to dust. The 1 μm diameter beam is moved by a linear motor system using servo tracks to give densities of about 25 000 TPI and data transfer rates in excess of 600 kbits mm^{-1}. By comparison, magnetic disks provide about 16 kbits mm^{-1}. Consequently, 2 Gbytes can be stored on a 12″ disk.

A bit cell consists of a blemish on the metallized surface. This blemish reflects light in a different manner to the rest of the surface. The data are therefore read by detecting the presence or absence of a blemish using the reflected light from a low-power laser directed onto the surface of the disk. Apart from this difference, optical disks are very similar to magnetic disks in the way they work (*e.g.*, track access, cyclic access). The form the blemish takes depends on the type of disk. For CD-ROMs/OROMs the blemish consists of pits pressed into the surface, for WORM, pits burnt into the disk or bubbles raised on the surface by heating, and for erasables, areas of different crystal structure.

One major difference between magnetic- and optical-disk drives is that in the optical system the head can be positioned several millimetres from the surface of the disk; 'head crashes' are therefore very unlikely. Optical systems are insensitive to surface contamination helping to make them 'user friendly'.

Data are encoded into the series of pits and lands using a process called EFM (eight-to-fourteen modulation). The pits and lands on the surface of the disk do not correspond to 1s and 0s; it is the transition between the pits and the land (*i.e.*, pit to land, or land to pit) that represents 1. The length of the pit or land indicates the number of 0s.

3.7.1 CD-ROM and OROM

The CD-ROM is exactly the same as the audio compact disk and hence has the following advantages: mass production is relatively easy, copies of the information on the disk being made by pressing the disks from a master (quicker than having to read the information from tape for each copy); the production methods are well established; and as the drives to read the disks are essentially the same as audio CD drives, they are relatively cheap and use proven technology.

Binary data in the centre of a disk will travel past a fixed point slower than a pit on the outer circumference of the disk. This is compensated for in magnetic disks by varying the space between the bits. The data can then be scanned at the same rate by a rotating disk at a fixed speed. This is called *constant angular velocity* (CAV). Direct addressing is then by sector and track. However, this is inefficient, as the storage capacity of the outer tracks is underutilized. The CAV method is not used for CD-ROMs, instead data are evenly packed and the disk is rotated at a variable speed. The data is read by the laser at a *constant linear velocity* (CLV). Direct addressing is then by units of 0 to 59 min, 0 to 59 s, and 0 to 74 blocks. The data are stored in pits pressed into the surface of the disk during production. The pits are 1 μm in diameter with a packing density of about 16 000 per inch, compared with 200 per inch on a hard magnetic disk. The data are read from the disk by directing light onto it from a low-powered laser that is in the reading head. It moves radially over the surface of the disk. The light reflected back from the disk is affected by the pits and the modulation of the light represents the digital data. Not all the pits contain 'data'. Data are arranged sequentially along a single spiral track. The use of CLV makes random access more difficult as it is necessary to move the head to the desired area, modifying the rotational speed, reading the address, and then making the necessary adjustments to locate the specific sector.

The OROM is similar to the CD-ROM, the differences being that instead of one spiral track the data are stored in concentric tracks as on magnetic disks, and that OROM uses CAV. In all other respects such as production, and methods of reading and head positioning, the OROM is identical to the CD-ROM.

The advantages of CD-ROM include high-storage capacity, simple and inexpensive mass production of disks, removable disks, and robust medium. The disadvantages include its read only ability, therefore no means for update, and its slow access time. An average access time for an optical disk may be from 100 ms upwards, compared with around 20 ms for a magnetic drive.

3.7.2 WORM

Write once read many optical disks allow data to be written to the disk. Once written, the data are indelible, and cannot be removed or overwritten. The data are read in the same manner as for OROM and CD-ROM. The disks for WORM drives are physically very similar to those of CD-ROMs and OROMs and consist of a metallized disk between glass disks for protection. There are different methods of writing to the disk. They all use a high-powered laser that produces the bit pattern on the preformatted track. The first method uses a thin metal film, usually tellurium. When the laser beam is focused on the film it melts a small area, producing a tiny hole approximately 1 μm in diameter. However, problems of oxidization and flaking around the hole can occur. In a second method, a sensitive polymer is overlaid with a metal film. When the laser is focused on the polymer it causes a blister to form. The metal is not vaporized. In a third method, the laser changes the characteristics of the bimetallic surface resulting in different reflective properties.

The blister method has the advantage of creating a smoother blemish on the surface. The pits are burned in, and so tend to have ragged edges. The blister method therefore suffers less from 'surface noise' due to ragged edges during the reading stage. The blister and pit are just as small as the preformed pits of the CD or OROMS and as densely packed.

3.7.3 Eraseable Optical Disk

Eraseable disks allow data to be written, read, deleted, and overwritten on the disk. The basic operation of the drives and the disk characteristics are the same as those of the WORM, CD-ROM, and OROM. The disks are blank, but preformatted (*i.e.*, grooved), and there are different methods used to read and write.

One method relies on magnetic optics (MO). The drive is constructed with a heat-resistant magnetic layer under the protective glass. At high temperatures, magnetism is lost at a certain point (the Curie point). A magnetic head next to the laser causes the material to take a particular magnetic polarity that reflects light differently, hence allowing binary reading. The magnetism on these disks is more resistant to external magnetic fields than standard magnetic media. The data are erased from the disk by reheating and applying a reverse magnetic field to return the polarity. One disadvantage of MO is the need for two passes to overwrite, one to erase and the other to record. However, development continues to address this problem.

A second method uses crystal morphology. Certain materials can be switched from a crystalline state (which is reflective) to an amorphous state (which is non-reflective) and back again by the use of a high-power laser. To write to the disk, the high-powered laser is used to heat discrete 1 μm areas (the bit cell) to above the temperature at which the change of state occurs; the areas of different reflectivity produced are used to store the data. Erasure is achieved by reheating the areas and changing the state back again. Reading is achieved by the low-power laser as before.

A third method involves crystal-to-crystal transition. A particular silver–zinc alloy exhibits reversible colours depending on its thermal history. It is normally silver in colour, with a hexagonal crystal structure. If it is heated to over 300 °C and cooled rapidly, it turns pink, and takes on a square crystal structure. If this in turn is heated to over 100 °C and cooled slowly, it reverts to silver. Writing is achieved by changing the colour to pink, and erasure by returning the areas to silver. Reading is achieved as before, the two colours having different reflectivity.

A fourth method uses dye polymer technology. An organic coating is placed on the disk to produce two layers that absorb light at different frequencies. Heating a point on the lower layer using one frequency raises a bubble that decreases the reflectivity. Heating the upper layer flattens the bubble and therefore increases reflectivity. Hence the intensity modulation of a reflected light beam represents the data.

3.7.4 Applications

The extraordinary capacity of these systems will lead to many new applications. Optical disks provide a storage medium particularly suitable for archiving, and have great possibilities in image processing and storage, *e.g.*, of medical records including digitized *X*-ray photographs. They are also likely to be applied to large databases. The WORM disk could be used for back-up of conventional magnetic media. Publishers will turn to CD-ROM increasingly, rather than offer on-line services. Where large reference databases are concerned, it is therefore possible that the effect on telecommunications will be a negative one. Suppliers of on-line services have line costs as high as 80% of their total costs. One disk can hold more information than a 1200 bits per second (bps) modem could deliver in a week.

3.8 DEVELOPMENTS IN MEMORY TECHNOLOGY

There are inherent disadvantages in the use of hard-disk drives, namely, they are susceptible to vibration and heat. Further, they have large power requirements.

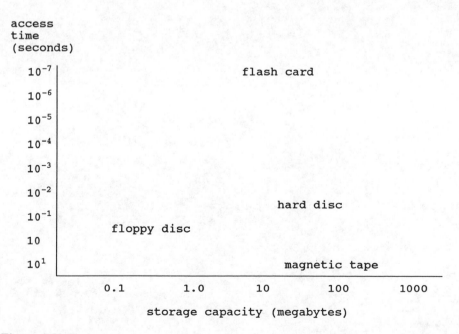

Figure 3.31 *Flash memory*

In response to demands to address these issues, hard-disk manufacturers continually improve disk performance. One of the smallest disks available is a 2.5″ disk capable of storing 120 Mbytes in a complete package of 65 × 100 × 20 mm. The access time is about 20 ms with active power consumption of 3 W and standby power consumption of about 500 mW. The disk can withstand a 5G shock and weighs 150 g. This can be contrasted with the new *flash memory card.* Flash memory devices are erasable and non-volatile. Data can be erased or recorded whilst still in the computer; however, at present this must be done in blocks of kilobytes. Flash cards are of comparable length and width to the 2.5″ disk package but are a mere 3 mm thick. Currently such a device can hold only 20 Mbytes, but development continues. Notably, such devices have access times of the order of 200 ns. They weigh less than 30 g and can withstand an impact of 50G. The active power consumption is 250 mW, but standby power is only 5 mW. It should be noted that flash memory can only be erased a limited number of times (hundreds of thousands of times) and that the cost exceeds that of the disk drive; however, the potential of this type of device is considerable (Figure 3.31).

CHAPTER 4

Operating Systems

A computer can be regarded as a collection of resources including both hardware (CPU, memory, peripherals, *etc.*) and software (word processors, databases, *etc.*). All of these resources must be managed in a controlled manner; this is done by the operating system. An operating system is a program for the efficient, safe, and simple operation of a computer (Figure 4.1). As such it should provide the following: an easy-to-use user interface; facilities for users with different requirements (casual users, programmers, computer operators, computer supervisors, *etc.*); utilities for program creation and execution (*e.g.*, text editors, compilers); simple access to software packages (*e.g.*, databases, spreadsheets); access to different types of I/O devices by means of simple commands; data security; and error trapping and control.

It is important to realize that the operating system is itself an 'overhead' requiring memory and CPU time. The more complex the operating system, the more demanding it is in memory requirements and CPU time. The operating system can therefore be considered as a resource manager. This is highlighted when one considers operating systems that execute more than one program 'apparently at the same time'—*multiprogramming*. This can be contrasted with *uniprogramming* where the computer executes only one program at a time. Uniprogramming is also called sequential programming.

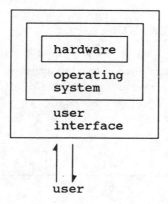

Figure 4.1 *The operating system shell*

89

4.1 MULTIPROGRAMMING/MULTITASKING

An operating system with multiprogramming/multitasking capability is able to manage several user programs simultaneously. Multiprogramming is the technique by which the processor and the rest of the computer resources are kept as busy as possible. A multiuser operating system allows several users to work on the same computer independently of one another provided each user has access to a terminal (keyboard and monitor). The CPU works at electronic switching speeds and, therefore, at orders of magnitude faster than electromechanical input devices and the response time of human operators. Assuming the time taken to read the data from one keyboard is 0.001 s, the time taken to process that data is 0.0001 s, and the time to output the result to the screen is 0.001 s.

Read one keyboard	0.0010 s
Process keyboard data	0.0001 s
Write result to screen	0.0010 s
Total	0.0021 s

$$\text{Percentage CPU utilization} = \frac{0.0001}{0.0021} = 4.8\%$$

Clearly, in a uniprogramming system the CPU does nothing for most of the time, and is in fact idle for about 95% of the time (Figure 4.2)!

In multiprogramming, the memory holds more than one user program. Consequently, when the program currently being executed is held up whilst waiting for a response from a peripheral, such as a user on a keyboard, the processor can switch to another program. This principle can be extended so that the memory holds many programs. The over-all result is an increase in CPU utilization (Figure 4.3). Each user program has a 'time slice'; this varies, but is usually about 50 ms. In a multiprogramming system there are algorithms that prevent a single user from taking up all the CPU time.

4.2 RESOURCE UTILIZATION

As we have perhaps gathered from our reading so far, terminology in computing is variable, and sometimes inconsistent and inaccurate. The term *job* is often used instead of program. The two main types of job are CPU bound requiring a lot of CPU time and little I/O (*e.g.*, calculations with complex algorithms), and I/O bound requiring little CPU time but a lot of I/O (*e.g.*, printing a very large

Figure 4.2 *Uniprogramming*

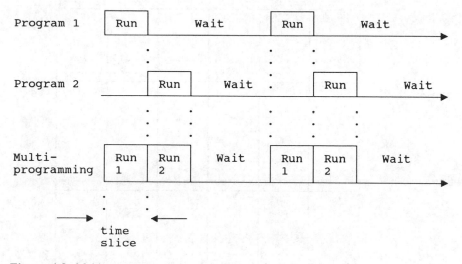

Figure 4.3 *Multiprogramming*

report). Consider a situation where there are three jobs to be processed (Table 4.1). The computer has 16 Mbytes of primary memory available for the programs. The results can be plotted on an activity histogram.

The uniprogram utilization histogram is shown in Figure 4.4. Given the relative speed of the CPU, this mixture of CPU and I/O bound jobs allows several jobs to run at the same time. As we will see shortly, the CPU works for a short time on each job in turn. The multiprogram utilization histogram is shown in Figure 4.5. Tabulating the results we can clearly see that not only are the

Table 4.1 *Job types*

Job type	CPU utilization/ %	Memory requirement	Disk activity/ %	Terminal activity/ %	Printer activity/ %	Duration/ min
1, CPU bound	50	8M	0	0	0	5
2, I/O bound	25	4M	25	50	0	15
3, I/O bound	25	2M	50	0	100	10

Table 4.2 *Percentage utilizations*

	Resource/% utilization						
Operating system	CPU	Memory	Disc	Terminal	Printer	Total time taken/min	Throughput/ jobs per hour
Uniprogramming	30	25	25	25	33	30	6
Multiprogramming	66	50	60	50	66	15	12

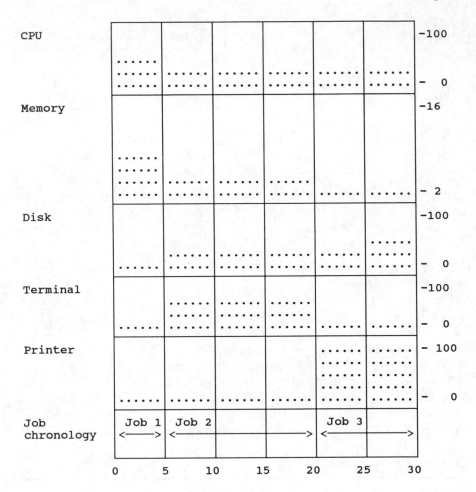

Figure 4.4 *Uniprogramming utilization histogram*

resource utilizations higher, but also the throughput (jobs per hour) is higher (Table 4.2). A cost accountant's dream! In multiprogramming, the main memory and CPU is shared between three (or more) different jobs at any one time. We must now consider how this can be done.

4.3 CENTRAL PROCESSING UNIT MANAGEMENT

I used to work on free radicals in the solid state. The standard tool for this type of work is an electron spin resonance spectrophotometer (ESR spectrophotometer). This is a large, expensive piece of equipment and we only had one that was worth using. There were many scientists all wanting to complete their work, but at any given time only one scientist could use the equipment. Who will use the instrument next? (Does everyone have equal priority or are there levels of import-

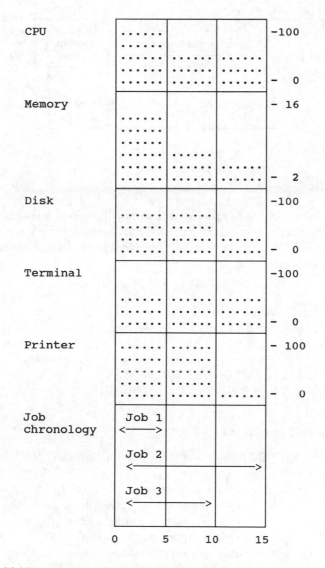

Figure 4.5 *Multiprogramming utilization histogram*

ance?) How long can an individual use the instrument for at a given time? (Until they have finished, until someone more important comes along, or up to a pre-defined time period?) These were the issues that we had to address.

In order to resolve the problem of infinite demand, but limited supply, we had a booking system whereby each scientist could book time on the spectrophoto-meter. There was an equal priority system, with the exception of the laboratory head who could use the instrument when he wanted, for as long as he wanted. Fortunately he rarely used it! This resulted in scientists existing in certain states. At any given instant there were some wishing to use the instrument, and some not

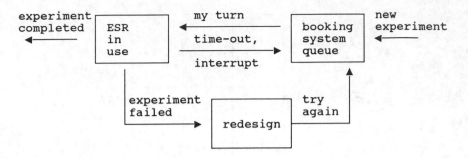

Figure 4.6 *Scientist states*

ready to use it (the previous experiment had failed, they were calculating their results, planning their next experiment, *etc.*) but wanting to use it later. But only one person was using the spectrophotometer. Certain events were the triggers for any changes in state (Figure 4.6).

Scientist States

ESR in use: one scientist uses the ESR spectrophotometer.

Booking system: many scientists want to use the ESR spectrophotometer. Their names are in a booking system queue.

Redesign: scientist not yet ready to use the spectrophotometer. There could be several scientists in this state.

Scientist State Transitions

My turn: scientist given control of the spectrophotometer. The laboratory head takes control.

Timeout/Interrupt: spectrophotometer control given up by the scientist. They have run out of time. Alternatively, the laboratory head wishes to conduct an experiment and the current occupant is interrupted.

Experiment failed: scientist unable to continue using the spectrophotometer—'back to the drawing board'

Try again: scientist now ready to use the spectrophotometer: book another time slot.

4.3.1 Context Switching and Interrupt Processing

When time 'runs out' for the current user of the spectrophotometer, the work to date must be concluded. This involves finishing the current experimental procedure, freezing the status of the experiment, *i.e.*, recording all instrumentation settings for reference next time (*e.g.*, magnetic field strength, modulation frequency), and logging all data in the experimental report.

In multiprogramming, a number of programs (analogous to our scientists) are

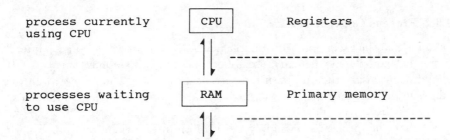

Figure 4.7 *Process context switching*

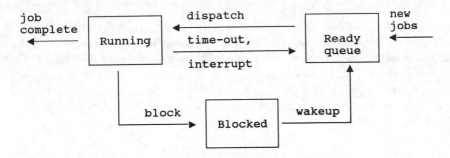

Figure 4.8 *Process state transitions*

resident in the main memory at the same time, and the CPU (analogous to our ESR spectrophotometer) is shared between them (Figure 4.7). Recall that only programs in the main memory can be executed by the CPU. A *process* can be defined as a program in execution. Processes can exist in different *states* and the changes between these states are called *state transitions* (Figure 4.8).

Process States

Running: process has the CPU.

Ready: process is able to use the CPU. There could be several processes in the 'ready' queue.

Blocked: process currently unable to use the CPU. There could be several processes in the 'blocked' state.

Process State Transitions

Dispatch: process given/takes control of the CPU.

Timeout/interrupt: the CPU control given up/taken away from the process.

Block: process unable to continue using the CPU (*e.g.*, waiting for data input).

Wakeup: process now able to use the CPU.

4.3.2 Process Control Blocks

Imagine that the execution of our simple machine code program from Chapter 2 is only half completed when an interrupt occurs or the process has 'timed out'. It would be necessary then to finish the current instruction, and to record all the values in the registers and the intermediate calculation values, *etc.*, obtained so far, *i.e.*, to 'freeze the status'.

So that the process can simply re-establish itself when necessary, each process needs information about itself (sometimes called the environment). This information is held in process control blocks (PCBs), also called task control blocks (TCBs). Each process has its own PCB and typically holds the following data: process identity (each process has a unique identifier); process priority (priority level of the process); process state (current state of process, *e.g.*, blocked); program counter (program counter value); pointers to allocated resources; and context data or the register save area. In effect, the process environment is stored on the PCB. The current job may be blocked because data input or output is required. Each input device may have a queue of processes requiring data I/O. The queue will be a list of PCBs (Figure 4.9).

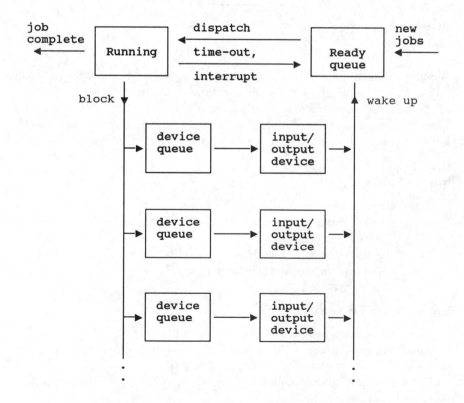

Figure 4.9 *Process queues*

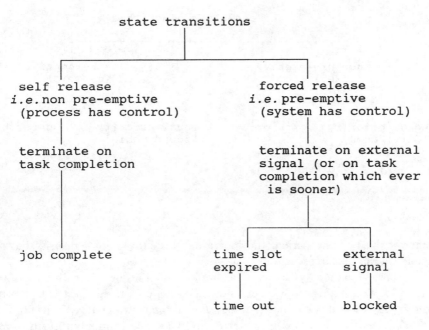

Figure 4.10 *State transitions*

4.3.3 Pre-emption

State transitions can be classified according to the method of release of the CPU resource.

In *non-pre-emptive* state transitions, the process has control of when it wishes to release the CPU. The process would normally release the CPU on completion of the task. In the case of our ESR instrument the user would have control for as long as they wanted. The current user releases control (Figure 4.10).

For *pre-emptive* state transitions, the operating system and not the process controls when the CPU will be released. The process is forced to release the CPU by a signal from the operating system. This external signal can be either that a more important process must be given control of the CPU or that the time allocated to the current process has expired. In the case of the ESR instrument the user has control for a predefined period or until a scientist with higher priority comes along. Control is taken away from the current user.

4.3.4 Scheduling

Let us consider the ESR spectrophotometer further. As we have seen there are two main types of algorithm controlling instrument access: non-pre-emptive and pre-emptive. Consider first the non-pre-emptive case. This means that at the start of each day (or week) the booking schedule is completed. There are different ways of organizing this (Figure 4.11). The equal priority 'first come, first served' (FCFS) basis, for example, suits the early risers. Alternatively, at the

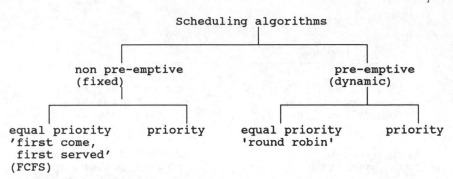

Figure 4.11 *Scheduling algorithms*

start of the booking period, the individual with the highest priority has first choice of the booking schedule.

A more dynamic system can be implemented using pre-emptive control. Again, there are different ways of doing this. In equal priority, also called 'round robin', each scientist has an equal *time slice* on the instrument. At the end of the time slice control is passed to someone else. Control eventually returns to each scientist. In the priority system, the current user can be interrupted if someone comes along with a higher priority project. It is they who take control.

Some of the algorithms may lead to '*starvation*'. Starvation is when a process never gets executed. This is especially the case in pre-emptive scheduling. Round robin is a scheduling algorithm that is particularly well suited to multiuser, interactive data processing. This ensures that all users get reasonable time on the CPU. Each job is given equal priority and has an equal *time slice* on the CPU (Figure 4.12). At the end of the time slice, control is passed to the next job. Control eventually returns to each job in turn. The important thing to note is that the CPU is working very fast—so fast that, if the system is not overloaded, the impression is given that each user has sole use of the computer. One disadvantage is that there is a context-switching overhead when the jobs are switched. However, this type of scheduling algorithm is not suitable to real-time

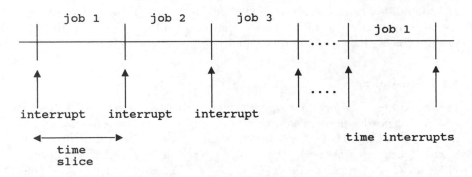

Figure 4.12 *Time slicing*

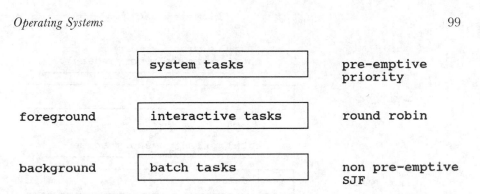

Figure 4.13 *Foreground and background jobs*

applications in which the computer is controlling equipment such as valves, heaters, *etc.* for which there must be no delay in responding to events.

Some computer systems employ a hierarchy of process scheduling. The operating system itself is a process and must therefore have system tasks running. These could be controlled by pre-emptive scheduling. The multiuser tasks or *foreground* jobs could be scheduled using a round robin algorithm with any low priority of *background* jobs being run on a batch basis (Figure 4.13). The batch jobs are run whenever there is low demand for CPU time.

In multilevel queues, the jobs are assigned to certain queues. There is no movement between queues. However *multilevel feedback* queues allow jobs to move between queues. Jobs are then automatically sorted according to CPU utilization. A job will enter the first queue (0), which allocates a quantum of say 10 ms. If this is insufficient to complete the job, then that job is moved to the tail of the next queue (1). If queue 0 is empty, then the jobs in queue 1 are processed with a quantum time of say 20 ms. There may be several layers of queues.

4.3.5 Process Schedulers

The term process scheduler is sometimes given to the operating system software that is responsible for selecting the next job to run. The co-ordinator then swaps the jobs

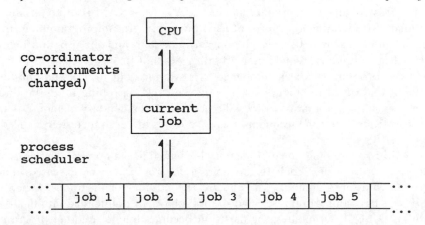

Figure 4.14 *Process schedulers*

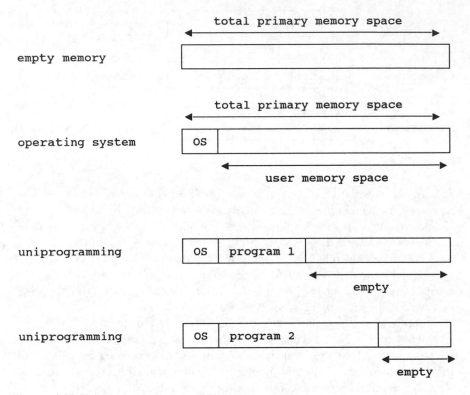

Figure 4.15 *Uniprogramming memory utilization*

(Figure 4.14). We have seen how the CPU can be shared; let us now examine how the primary memory can hold many programs simultaneously.

4.4 MEMORY MANAGEMENT

Let us start with an empty primary memory (IAS, *i.e.*, a RAM IC). To this must be added our operating system (or at least part of it). In uniprogramming, the main memory is shared between the operating system and the program being executed, and there is a single user program resident in the primary memory. When this program has been completed it is swapped out to disk and another program is swapped in (Figure 4.15). However, disk-swapping time, the time taken to exchange programs between the primary and secondary memories is an overhead. Recall that the secondary memory works at very much slower speeds (Figure 4.16).

In multiprogramming, the user memory is divided into a number of regions or *partitions* (Figure 4.17). Each region holds one program to be executed. The degree of multiprogramming is bound by the number of partitions. When a partition is free, a program is selected from a job queue and loaded into the free partition. When it terminates, the partition becomes free for another program. Multiprogramming is much more dynamic, with programs continually being

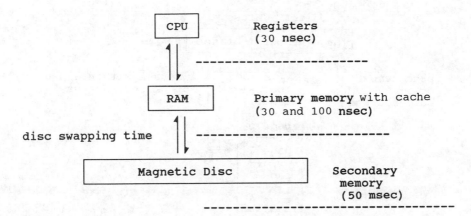

Figure 4.16 *Disc swapping time*

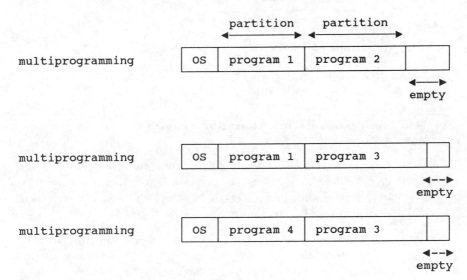

Figure 4.17 *Multiprogramming memory utilization*

swapped into and out of the primary memory. Program 2 is completed, swapped out to the secondary memory, and program 3 is moved from the secondary memory into the primary memory. Program 1 is completed, swapped out to the secondary memory, and program 4 is moved from the secondary memory into the primary memory. Unused memory space is called fragmentation. Multiprogramming requires effective memory management, and particularly, *memory protection*. A program in one partition must not be allowed to access memory locations in another partition. There are different types of memory management scheme. With the early types of memory management the entire program had to be in one partition, *i.e.*, in a contiguous memory space.

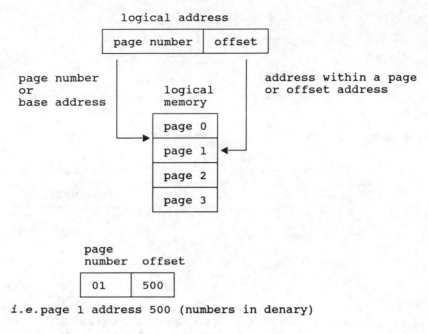

Figure 4.18 *Logical address*

4.4.1 Non-contiguous Memory Allocation—Paging

The early techniques of memory management suffered from fragmentation and the need for a contiguous memory space. These problems have been addressed with *paging*. Paging allows a program to be non-contiguous within the memory, thereby allowing a program to be allocated memory wherever it is available. Also, fragmentation is kept to a minimum.

4.4.1.1 Logical Addressing and Logical Memory. The addresses used by the program in paging have two fields, the *page number* and the *offset*. We now have a *logical address space* that is divided into *pages* with each page being, for example, 1K long. The page number defines the logical memory page, and the offset defines the address within that page (Figure 4.18).

4.4.1.2 Primary Memory Paging. The primary memory is divided into *frames*, which are the same size as pages. The addresses used have two fields, the *frame number* and the *offset*. The frame number defines the frame to be used, and the offset defines the address within that frame (Figure 4.19).

4.4.1.3 Page Tables. When a program is loaded into the memory, its pages are loaded into any available frames, and the *page table* is used to translate from the logical memory pages to the primary memory frames. When the program uses an address, the page number is used as an index into the page table. The page table contains the associated frame addresses of primary memory. The offset is combined with the associated frame address to form the new address. In our example,

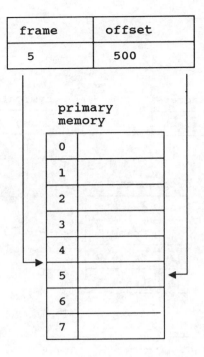

Figure 4.19 *Physical address*

the logical page number 1 is found in page frame number 5. Page frames are the same size as logical pages, hence the offset remains the same.

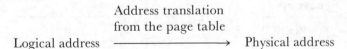

The consequence is that the logical memory for a program is not contiguous in the primary memory, but dispersed, although it appears to the program to be contiguous (*i.e.*, artificial contiguity) (Figure 4.20). In operation, each program has a page table. Internal fragmentation still occurs, but with small page sizes, this is minimized.

4.4.1.4 Shared Pages. Paging gives the possibility of sharing code. This is of considerable importance in time-sharing systems. Consider a multiuser system where several people wish to use the same software, *e.g.*, a word processor. Normally we would need as many copies of the word processor as there are users. If the word processor is particularly big this would be extremely wasteful of resources. The use of *re-entrant code* allows a single copy of such software to be shared. In the following example Figure 4.21, two users share the code. The re-entrant code is also referred to as common, pure, executable, and even non-self-modifying code. The important characteristic is that it never changes during execution thereby allowing many users to execute the code at the same

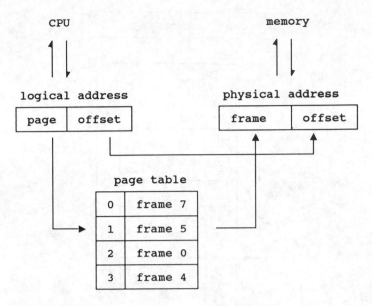

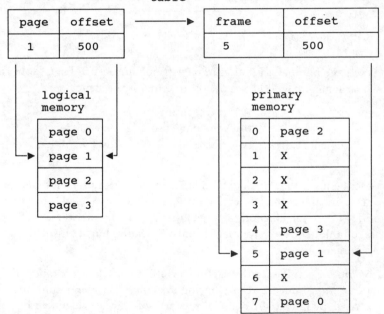

X : pages used by other programs

Figure 4.20 *Paging*

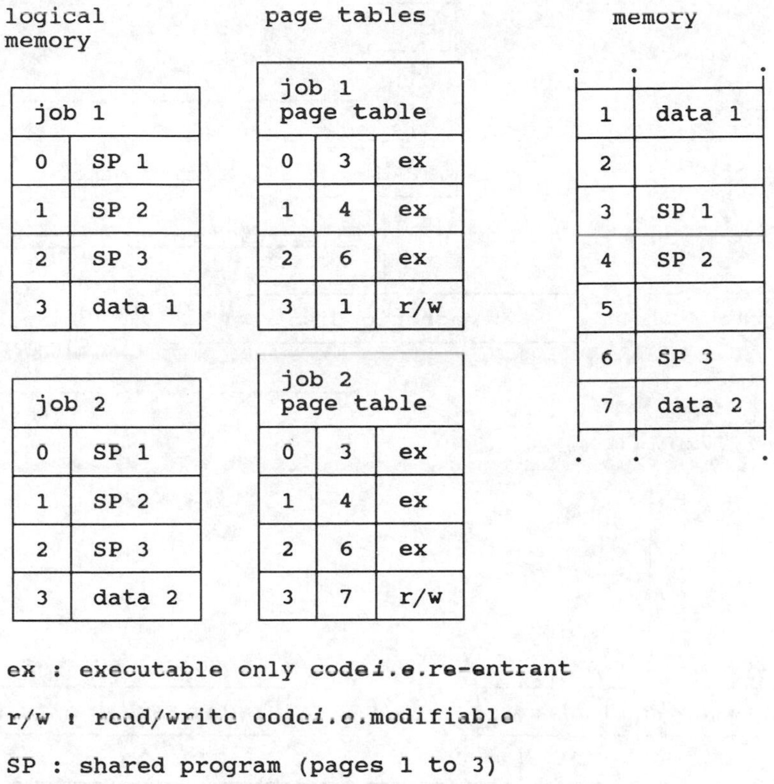

| logical memory | page tables | memory |

job 1		job 1 page table					
0	SP 1	0	3	ex		1	data 1
1	SP 2	1	4	ex		2	
2	SP 3	2	6	ex		3	SP 1
3	data 1	3	1	r/w		4	SP 2

						5	
job 2		job 2 page table				6	SP 3
0	SP 1	0	3	ex		7	data 2
1	SP 2	1	4	ex			
2	SP 3	2	6	ex			
3	data 2	3	7	r/w			

ex : executable only code*i.e.*re-entrant

r/w : read/write code*i.e.*modifiable

SP : shared program (pages 1 to 3)

Figure 4.21 *Shared pages*

time. Note that each user process has its own copies of data storage and registers to hold the read/write data.

4.4.2 Virtual Memory

In multiprogramming there are many programs in the primary or main memory at the same time. However, they all require that the entire program is in the primary memory, either contiguously or in pages. Virtual memory allows the execution of programs that may not be completely in the primary memory, the consequence being that *the number of active programs can exceed the capacity of the primary memory store* (Figure 4.22). Further, the job can be larger than the physical size of the primary memory. The main difference between paging and virtual memory is that the page table holds disk addresses. The *virtual page* is used to access the page table. If the first bit in the associated page table is 0 this is a page fault, the required page is not in the primary memory and the next field supplies the corresponding disk address. If, however, the first bit is 1, then the disk address field is ignored and the following field supplies the page frame number (Figure 4.23). The page table shows that virtual pages 1 and 2 are in the primary

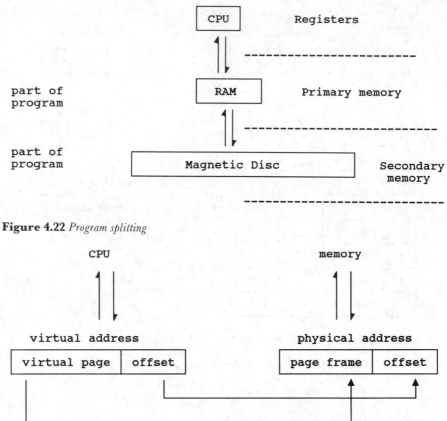

Figure 4.22 *Program splitting*

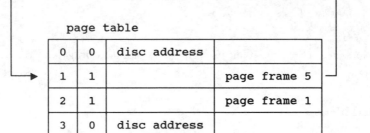

Figure 4.23 *Virtual page table*

memory but pages 0 and 3 are stored on disk. From the page table, virtual page 1 becomes page frame 5 with the same offset value (Figure 4.24). The virtual address has now been translated to a physical address in primary memory. From the page table, virtual page 0 is not in primary memory. When an attempt is made by the CPU to address this page there is a *page fault*. The desired page is held in secondary memory (disk). The disk address field in the page table holds the secondary memory address—cylinder number, surface number, sector number. The virtual address has now been translated to a physical address in the primary, *i.e.*, virtual memory (Figure 4.25). The page that is currently in frame 3

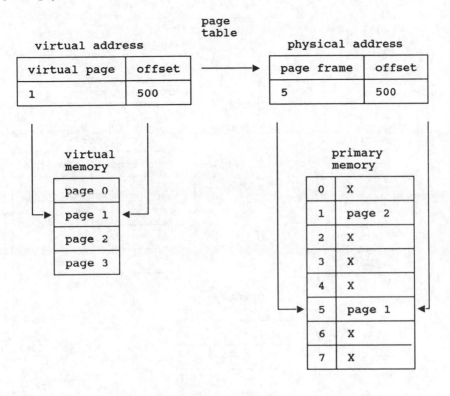

X pages used by other programs

Figure 4.24 *Virtual to physical address translation*

is copied back to the secondary memory and then this space is used. After a page fault, it is necessary for the operating system to read in the required page from the disk and for the page table to be updated accordingly (Figure 4.26). This is called *demand paging*. The analogy is the feeding algorithm for babies—when the child cries, feed it (as opposed to predefined feeding periods). Consequently, a page (sometimes called the victim) will have to be removed from the primary memory and placed on disk. Various algorithms exist for deciding which page in the primary memory will be the victim, but they will not be discussed here.

4.5 FILES

Records, files, and file processing will be considered in more detail later. For the moment, let us briefly define a record as an indexed collection of related data, and a file as a named collection of records. The simplest analogy is a card index or filing cabinet (Figure 4.27).

The individual data items (records) within the file may be manipulated by the following operations: *read*, *i.e.*, the data is read from the file (analogous to reading

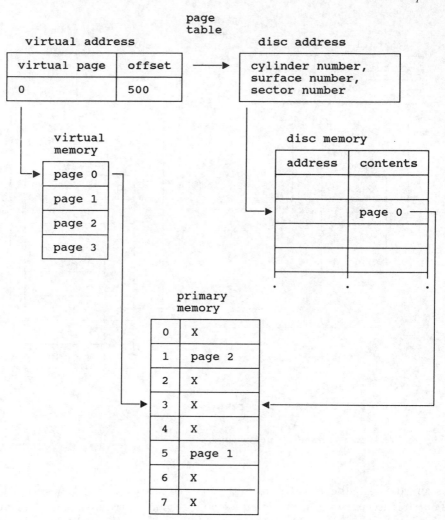

Figure 4.25 *Virtual memory*

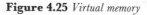

page table

0	1		page frame 3
1	1		page frame 5
2	1		page frame 1
3	0	disc address	

Figure 4.26 *Updated virtual page table*

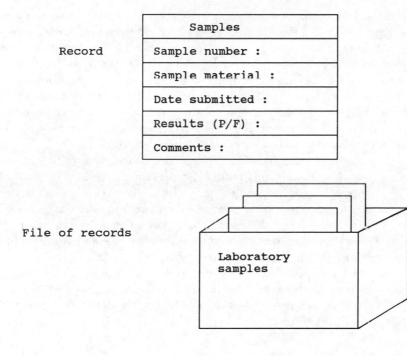

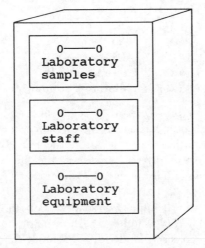

Figure 4.27 *Record and file*

a record); *write, i.e.*, the data is written to the file (analogous to writing data onto a record); *insert, i.e.*, add a new data item to the file (analogous to putting a new record in the file); and *delete, i.e.*, remove a data item from the file (analogous to removing a record from the file).

Files normally reside on secondary storage devices. We may have a collection of files (Figure 4.28), which may be manipulated as a unit by the following

Figure 4.28 *Collection of files*

operations: *open*, *i.e.*, data in the file may be referenced (analogous to opening a filing cabinet); *close*, *i.e.*, data in the file may not be referenced (analogous to closing the filing cabinet); *create*, *i.e.*, create a new file (analogous to getting access to another filing cabinet drawer); *delete*, *i.e.*, delete a file (analogous to emptying the contents of the filing cabinet drawer in the trash can); *copy*, *i.e.*, copy the file contents to another file name (analogous to transferring the contents of one drawer to another); *rename*, *i.e.*, change the name of the file (analogous to changing the name on the drawer); and *list*, *i.e.*, display the file contents (analogous to removing the contents of the drawer for access).

Files store data and programs. Other properties associated with files include time of creation, length, and storage time. At the machine level there are only bits. The operating system provides the user with a schematic view of this data and the relationships between the data items. The operating system takes the bits and the bytes from the storage device to present a file of records to the user. The operating system knows the structure of the file and maps the abstract concept of a file onto mass storage devices such as magnetic tape and disk.

Typically, files are organized into directories for ease of use. Files may be multiple access and in this case controls are needed to address such issues as privacy. We will consider records and files in greater detail in Chapter 7, Introduction to Data Processing.

4.5.1 Director Systems

A directory is a method for organizing many files. It may span several device boundaries, but this is hidden from the user.

Tree-structured directories allow users to create their own subdirectories and to organize their files accordingly. The tree is said to have a root directory, with every file in the system having a unique path name. A path name is the path from the root, through all the subdirectories, to a specific file (Figure 4.29).

Filing cabinet 1 contains three drawers.

Drawer 1: Laboratory samples

92

93

94

Drawer 2: Laboratory staff

Drawer 3: Laboratory equipment

A directory, or subdirectory, contains a set of files and/or subdirectories. In operating systems, each user has a current directory, their own filing cabinet. This contains the information most relevant to the user. References to files result in a search of this current directory. Should the file not be in the current directory, the user must either change the directory or specify the path name appropriately.

Path names can be either complete or relative. A path name to a file is a text string that identifies a file by specifying a path through the directory structure to the file. Syntactically it consists of individual file name elements separated by the

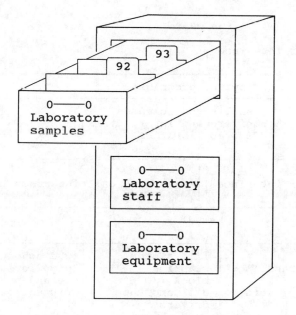

Figure 4.29 *File directory*

slash character. For example, /laboratory samples/92/July. The first slash indicates the root of the directory tree, called the root directory, in our case, the filing cabinet. The next element, laboratory samples, is a subdirectory of the root, 92 is a subdirectory of laboratory samples, and July is a file or a directory in the directory 92. With this syntax, it is not possible to determine if 92 is a file or a directory.

Absolute path names start at the root of the file system and are distinguished by the slash at the beginning of the path name. Relative path names start at the current directory.

4.6 CASE STUDIES

Three operating systems are now going to be considered, DOS, OS/2, and UNIX. Certainly each is a study in itself. The case studies are meant to illustrate some important design considerations and thereby to address the following questions. What type of CPU scheduling is employed? Does the operating system use virtual memory? What applications are best suited to this operating system?

4.6.1 UNIX

UNIX was designed to be a time-sharing system for supporting multiple processes, *i.e.*, it is suitable for a community of users. It consists of two separate parts, the *kernel* and the *systems* programs (Figure 4.30). The kernel provides the file system, CPU scheduling, memory management, *etc.* Systems programs use the kernel to provide useful functions such as file manipulation.

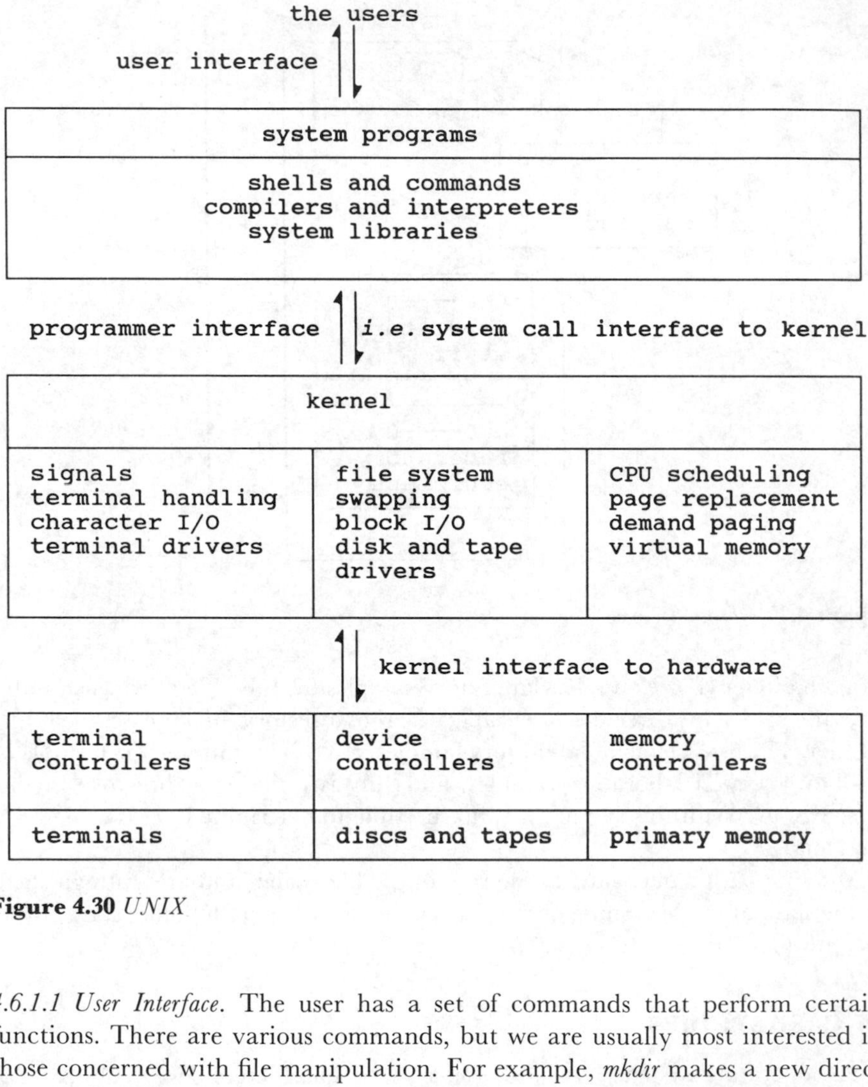

```
                          the users
                             ↑
       user interface        ‖
                             ↓
  ┌─────────────────────────────────────────────────────────────┐
  │                      system programs                         │
  ├─────────────────────────────────────────────────────────────┤
  │                   shells and commands                        │
  │                compilers and interpreters                    │
  │                    system libraries                          │
  └─────────────────────────────────────────────────────────────┘
                             ↑
  programmer interface       ‖  i.e. system call interface to kernel
                             ↓
  ┌─────────────────────────────────────────────────────────────┐
  │                          kernel                              │
  ├─────────────────────┬─────────────────────┬─────────────────┤
  │ signals             │ file system         │ CPU scheduling  │
  │ terminal handling   │ swapping            │ page replacement│
  │ character I/O       │ block I/O           │ demand paging   │
  │ terminal drivers    │ disk and tape       │ virtual memory  │
  │                     │ drivers             │                 │
  └─────────────────────┴─────────────────────┴─────────────────┘
                             ↑
                             ‖  kernel interface to hardware
                             ↓
  ┌─────────────────────┬─────────────────────┬─────────────────┐
  │ terminal            │ device              │ memory          │
  │ controllers         │ controllers         │ controllers     │
  ├─────────────────────┼─────────────────────┼─────────────────┤
  │ terminals           │ discs and tapes     │ primary memory  │
  └─────────────────────┴─────────────────────┴─────────────────┘
```

Figure 4.30 *UNIX*

4.6.1.1 User Interface. The user has a set of commands that perform certain functions. There are various commands, but we are usually most interested in those concerned with file manipulation. For example, *mkdir* makes a new directory, *rmdir* removes a directory, *cd* changes the current directory to another, and *more* displays the file contents on the screen one page at a time. Execution of these commands and others are carried out by the command-line interpreter.

4.6.1.2 Programmer Interface. A file in the UNIX system is a sequence of bytes. Different programs might expect certain file structures, but the UNIX kernel does not impose a structure on files. An example of a file structure is that of a text file, which consists of lines of ASCII characters separated by a single new-line character.

The files are organized in tree-structured directories, much as in the DOS. The directories are themselves files that contain information on how to find other files.

A path name to a file is a text string that identifies a file by specifying a path through the directory structure to the file. Syntactically it consists of individual file name elements separated by the slash charcter. For example, in /db/lims/ compex, the first slash indicates the root directory, the next element, db, is a subdirectory of the root, lims is a subdirectory of db, and compex is a file or a directory in the directory lims. With this syntax it is not possible to determine whether compex is a file or a directory. UNIX has both absolute and relative path names. Absolute path names start at the root of the file system and are distinguished by the slash at the beginning of the path name. Relative path names start at the current directory. The file name '.' in a directory is a hard link to the directory itself. The file name '..' is a hard link to the parent directory. Hardware devices have names in the file system. These device files or special files are known to the kernel as device interfaces.

4.6.1.3 Central Processing Unit Management. Processes in UNIX are represented by PCBs. The information in these control blocks is used by the kernel for process control and CPU scheduling. The PCBs contain everything that it is necessary to know about a process when it is swapped out (*i.e.*, process identification and scheduling information).

Scheduling of the CPU is designed to benefit interactive processes. Processes are given CPU time slices on a round robin basis, with a 0.1 s pre-emption time slice that takes priority into account. Every process has an associated scheduling priority. The larger the number, the lower the priority. The more CPU time a process accumulates, the lower its priority.

4.6.1.4 Memory Management. Decisions on which processes to swap in or out are made by the process scheduler. The scheduler is periodically activated to check for processes to be swapped. UNIX employs demand page virtual memory. The paging method minimizes fragmentation.

4.6.1.5 Input/Output. One of the primary functions of any operating system is to provide a consistent user interface independent of the underlying hardware. UNIX addresses this problem by isolating the I/O devices in the I/O system. Isolation of machine-specific code means that it is relatively easy to make the necessary modifications to the operating system thereby allowing it to run on different computers using different I/O devices. There are three main types of I/O device, block, character, and socket interface.

Block devices are secondary memory devices. They are characterized by fixed blocks (typically 512 bytes) that are addressable. The block driver isolates the specific details of the drive. Character devices include VDU terminals and line printers.

UNIX is a non-proprietary and vendor-independent operating system. Any vendor can purchase a source code licence and modify the operating system to run on their computer. Consequently, by the late 1980s, well over 100 different versions of UNIX existed. Some of the major systems include AT&T System V, Berkeley UNIX 4.3, and IBM AIX. The progressive importance of international standards has led to a movement towards standardizing UNIX. As such, UNIX is considered a critical component in *open operating systems technology*. The IEEE POSIX standard includes a standard UNIX interface definition and is based on

AT&T System V with popular enhancements. This standard also includes the X/Windows standard definition describing the base architecture that is used to support graphical user interfaces. Another major standards body is X/Open. This organization was originally a consortium of European vendors concerned with specifying a *common application environment* (CAE) based on both formal and *de facto* standards. Most major UNIX vendors belong to X/Open. The Open Systems Foundation (OSF), again a consortium of vendors, including IBM, DEC, and Hewlett Packard (HP), is concerned with defining a standard UNIX. The first UNIX implementation, OSF/1, uses the Motif graphical user interface based on X/Windows.

Features

Advantages
    Multitasking
    Multiuser

    16 Mbytes of linearly addressable
    memory
    MS-DOS applications run under
    some versions
    Good network capability
    Electronic mail
    Graphical user interface available
    (X/Windows)
    Open Systems

Disadvantages
    Large memory requirements
    Not suitable for real-time
    applications

Technical Specifications

Hardware requirements	
Processor	80386, 80486 or comparable
main memory	approximately 4 Mbytes
One hard disk	at least 40 Mbytes
Operating system	
Mode	protected
Memory requirement	minimum 4 Mbytes
Address space	16 Mbytes
Main console and up to 16	
additional terminals	
Number of concurrently	
executable programs	any number, usually 10–20
Response time	depends on system load
Interprocess communication	yes
Maximum program length	16 Mbytes

4.6.2 Disk Operating System (DOS)

DOS is the most widespread PC operating system and as such it is a *de facto* standard for use in commercial applications to which many software vendors and users conform. It is a single-user, single-tasking operating system. It was designed to allow one program at a time to use the CPU and other resources. Being sequential, DOS is the simplest type of operating system. It does not provide any separation between the operating system and the running machine. Both the system and the programs can access all facilities in the machine, including special instructions, ROM routines, and I/O ports controlling peripheral devices. Also, both the system and programs are executed using physical memory addresses and have the capability of altering each other. The DOS system is composed of the DOS kernel, which provides all the system and supervisor functions, and device drivers, which provide a software interface between the system and the hardware. A user shell allows the user to run programs and interact with the system. In the early versions of DOS this shell is a primitive line-oriented command processor. A simple graphical user interface is provided by DOS 4.0. Two types of device are supported by DOS, block devices (disks, diskette drives) and character devices (keyboards, printers, *etc.*).

The first edition of the DOS (1.0) for the IBM PC was released in 1981. It supported PCs with up to 256 Kbytes RAM and two 180 Kbytes floppy-disk drives, and it included a basic I/O system (BIOS) built into the ROM system. The primary data structure used by the DOS file system to map file blocks to diskette addresses was the FAT; the DOS file system became known as the FAT file system. The DOS 1.1 was released in 1982 to support PCs with 360 Kbyte floppy-disk drives. The PC/XT was introduced in 1983. This included a hard disk and a system board that allowed 640 Kbyte memory to be installed. The introduction of the hard disk led to a problem for the DOS 1.1, as the 360 Kbyte floppy-disk drives could contain a maximum of 64 files. This problem was solved with a hierarchical file-management system, much like UNIX. The next version of DOS also had to include an architecture for supporting different peripheral devices. This was solved by a user-installable program that interfaced between the DOS system and the applications to devices. These modules are known as device drivers. The PC/AT, based on the Intel 80286, was shipped in 1984. Several modifications to DOS 2.0 were required to support the AT hardware, and therefore DOS 3.0 was not released until 1984. The DOS 3.1 was released in 1985 to support PC (local area networks, LANs). Another upgrade was made in 1986 (DOS 3.2) to support 3.5″ disk drives (Table 4.3).

4.6.2.1 Microprocessors—an Overview. To be able to understand the background to the design of DOSs and of the OS/2, which will be considered next, it is necessary to have some knowledge of the Intel microprocessor family.

The Intel 8088 and 8086 are 16 bit general-purpose microprocessors. The 8088 and 8086 are identical, except that the 8088 has an 8 bit external Data Bus (DB) and the 8086 has a 16 bit external DB. The 8088 can address up to 1 Mbyte of

Table 4.3 *Evolution of the DOS*

Year	DOS version	Machine	Processor	Resident code	Characteristics
1981	1.0	PC	8088	4K	No hard disk; 64 files max
1983	2.0	PC/XT	8088	24K	Hard disk; subdirectories
1984	3.0	PC/AT	80286	36K	Hard disk; subdirectories
1985	3.1	PC/AT	80386	46K	Hard disk; subdirectories; networking
1987	3.3	PS/2	80X86		New line of PCs
1989	4.0	PS/2	80X86		Simple multitasking

physical memory. A physical location in the memory is addressed with two 16 bit values, the segment and the offset. The segment value notes the start of a 64 Kbyte region, and the offset is the number of bytes from the start of this region. Because of this segmented addressing scheme, the source code written for the 8088 is only portable to systems using exactly the same segment semantics and addressing scheme. The 8088 does not provide memory protection or I/O protection, and it is therefore not a suitable multitasking platform. The 8088 has only a small register set. This and the segmented memory model adds a level of complexity to the development of programs and tools to support the 8088.

The Intel 80286 was the first microprocessor to provide a protected environment. It has two modes of operation: *real mode* and *protected mode*. In real mode, the 80286 behaves like a fast 8088, and is fully compatible with the 8088. Real mode is the mode of operation defined by the characteristics of the Intel 80XX microprocessors. The 80XXX series were designed to provide additional features but to retain compatibility with the very large 80XX software base. Intel therefore incorporated an 80XX mode within the 80XXX series called the *real mode* or 80XX compatibility mode. The *protected mode* is normal mode for the 80XXX series and allows the processors to perform as designed with the associated memory addressing capabilities. This mode is incompatible with real mode. Software designed to run in real mode will not run in protected mode; it has to be rewritten. When the 80286 is running in protected mode, the processor provides an architecture that supports virtual addressing, memory protection, I/O protection, and access to up to 16 Mbyte of physical memory. The processor still uses a segmented memory model with a maximum segment size of 64 Kbyte. The protected mode of the 80286 uses virtual addressing, where a construct called a descriptor is used for mapping virtual addresses to physical addresses. Virtual addressing makes segment relocation transparent to applications using this addressing scheme. The 80286 protected mode memory model is different from the 8088 (real mode) memory model. With the 8088, there is a direct relation between the segment values and the real storage addresses of the segment.

The 80286 provides a protection model that allows an operating system to be isolated from user applications, to isolate user applications from one another, and to validate memory accesses. The 80286 has four privilege levels (0–3) that allow system code and data to be protected from less trusted code. A process can only access data that are at the same or at a less trusted privilege level. Whenever memory is referenced, the memory management unit hardware (MMU) on the 80286 is used to verify that the attempt satisfies the protection criteria.

Generally, the 80286 provides a suitable platform for a protected virtual memory multitasking environment. The maximum 64 Kbyte segment size is still a limitation, making programming difficult and the source code relatively non-portable. However, the 1 Mbyte barrier associated with the 8088 can be broken with the 80286. It is not feasible to make use of this for real-mode programs because of differences in segment semantics.

The Intel 80386 processor is used in some IBM PS/2 and AT systems. Like the 80286, the 80386 has a real mode and a protected mode. In real mode, the processor behaves like a very fast 8088, and is therefore compatible with all 8088 programs. In protected mode, the 80386 is compatible with all protected mode 80286 software. The 80386 has a third mode called *virtual* 8086 mode. This mode offers execution of 8086 programs in a protected environment. It enables system software to emulate an 8086 environment with a virtual machine. This provides the platform for concurrent execution of real mode 8086 applications in a protected environment. The *8086 virtual mode* is restricted to the 80386 and 80386SX and upwards processors and enables the user to run several protected real-mode sessions as one or more subtasks within the 386 chip. Multiple DOS programs can run as if they were running on their own single processor. Each virtual processor needs 1M of memory.

The 80386 register set is stretched to 32 bits to support 32 bit arithmetic, and has segments of up to 4 Gbytes, and a physical memory configuration of up to 4 Gbytes. Underneath the segmented memory architecture, the 80386 provides a

Table 4.4

Processor	*AB*/bits	*DB*/bits	*Coprocessor*
8088 (internal data registers, 16 bit)	20	8	8087
8086	20	16	8087
80286	24	16	80287
80386SX (internal data registers, 32 bit)	24	16	80387SX
80386DX	32	32	80387DX
80486SX	32	32	external
80486DX	32	32	built in
Pentium (80586)	32	64	

Table 4.5 *Intel addressing range*

Intel microprocessor	AB/bits	Segment size		
		Kbytes	Mbytes	Gbytes
8088/8026	20	1024	1	
80286	24	16384	16	
80386	32		4096	4

paged memory architecture. The 80386 exists in two versions, DX and SX. They are identical, except that the DX has a 32 bit external data/address path and the SX a 16/24 bit external data/address path. The 80386 provides the necessary functions for a protected 32 bit multitasking environment. The 32 bit model can be used for breaking the 64 Kbyte segment size of the 80286. The virtual 8086 mode is an important feature, allowing concurrent execution of 8086 programs in a protected environment.

The Intel 80486 processor is an 80386 compatible processor with several changes to improve performance. The processor has an 8 Kbyte on-chip cache for frequently used instructions. The cache is a small, but fast memory. The processor also features a built-in 80387 numeric coprocessor and achieves pipe-lined execution, which in most cases improves the performance. The 80486SX does not have the built in co-processor.

The 8086/8088, 80286, 80386, and 80486 microprocessors have real mode in common. This supports 16 bit execution and a memory divided into 64 Kbyte segments. Real mode is suitable for single-user, single-task, operating systems like the DOS. The 80286, 80386, and 80486 all provide another mode, protected mode, which is suitable for virtual memory, and program and system isolation. This mode also allows access to up to 16 Mbytes (80286) and 4 Gbytes (80386/ 80486) while providing a protected environment suitable for virtual memory multitasking environments such as the OS/2 or UNIX (Figure 4.31). Further, the 80386 and 80486 provide a virtual 8086 model, which allows multiple 8086 programs to run in a protected environment.

4.6.2.2 Limitations of DOS. As time went by, the DOS system and the 8088 environment proved to have some limitations. These were in the areas of memory management, I/O management, multitasking, and graphical user interface.

The 1 Mbyte address space of the 8088, which in 1980 seemed very large, was soon to become a major limitation for large DOS programs. Applications such as spreadsheets and databases allowed the user to create large volumes of data that needed to be in the physical memory to be processed. This memory limitation soon became known as 'the 640 Kbyte barrier', since only 640 Kbytes of the 8088 address space mapped RAM. The addresses from 640 Kbyte to 1 Mbyte in the PC were used to map the system ROM and the memory mapped I/O devices. In addition, the DOS device drivers used a portion of the 640 Kbyte address space. To relieve this problem, application programmers used an overlay scheme that

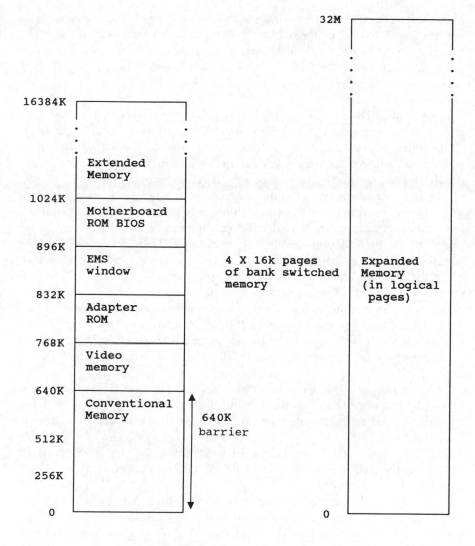

Figure 4.31 *Memory*

allowed a portion of program not currently needed to reside on a secondary storage device. The DOS system only provides a primitive memory manager, so the applications had to provide their own overlay management. The memory problem was made worse by the terminate and stay resident (TSR) modules. A TSR module is loaded like any other program, but stays resident in the memory after termination. Thereafter, the TSR module is accessed by a hardware or software interrupt. An example of a TSR module is a DOS print spooler that intercepts timer/printer interrupts to allow simultaneous queuing and printing of files. Terminate and stay resident modules can be unpredictable, because they have to allocate all the memory they need when they are loaded. These modules are not aware of one another's existence and resource requirements, and can therefore, in the worst case, halt the system. This is caused by the lack of memory protection features in the 8088. There is no distinction between the accessing system and the application memory, so applications can therefore modify one another or the system.

Another 8088/DOS problem is the lack of I/O control. Applications can freely read from or write to any I/O device without being granted access by the DOS. This can cause the system to stop or 'hang', or it can cause secondary storage data to be destroyed. Applications also have the capability of disabling interrupts. Again, this can cause the system to hang or behave unpredictably.

The DOS was designed to run one application at a time; it is a single-task or single-thread environment. Even in a simple environment like that, a certain level of multiprogramming is needed, for example, to operate a print-spooler to print a file in the background while editing another. There are no multitasking facilities in DOS, so applications have to provide these facilities themselves. Since the DOS is not re-entrant, only one program can use the DOS system services at a time. Competing applications and the TSR module can enter the system at the same time, and may therefore disrupt the system. Another reason for including multitasking facilities in the operating system kernel is the need for efficient use of CPU time. In DOS systems, applications typically use polling for I/O operations—a waste of valuable CPU time that could be used for other tasks.

DOS is equipped with a line oriented command processor. This means that new users must learn DOS commands before using the system. This is often found to be very complex and must therefore also be considered to be a DOS limitation.

4.6.2.3 Approaches to DOS limitations. As the limitations of the DOS operating system and the 8088 architecture were realized, different approaches were taken to relieve the worst problems.

Prior to, and during the development of the OS/2, Microsoft designed a graphical user interface (GUI) for the DOS environment called Windows. This is now the *de facto* standard for DOS systems. Windows is a window/icon based user interface with many similarities to the OS/2 Presentation Manager. It should be mentioned that Windows is not an integrated part of the operating system, unlike the OS/2 Presentation Manager of the GUI used by Apple Macintosh.

Expanded memory is memory that is not in the microprocessor's direct address space and as such is accessible via bank switching. This is a memory paging technique that provides small windows of memory (physical pages) through which blocks of expanded memory are used. This memory is lower down the memory hierarchy than primary memory. The expanded memory specification (EMS) allows four contiguous physical pages of 16K each to access up to 32M of expanded memory through the expanded memory manager (EMM). Lotus Intel and Microsoft (LIM) collaborated to devise the EMS. This memory is slower than normal primary memory. Systems with the 80386 processor do not need an expanded memory board due to the processor's memory management compatibilities that provide up to 4 Gbytes of possible extended memory as simulated expanded memory. This configuration is fast because all the work is done in the processor. The DOS expanded memory is one way of allowing DOS applications to access more than 640 Kbytes. Expanded memory is not a long-term solution to memory problems, as applications have to do their own memory management. However, programmers were able to solve the problems of the 640 Kbyte barrier in the short term (Figure 4.31).

The portion of memory past the first megabyte is called extended memory; this provides another way of relieving memory problems. Extended memory refers to memory that can be added to the PC above the 8088 1 Mbyte address space. Although no special bank-switching hardware is required, special memory-extending software is still needed.

DOS extenders allow applications to utilize the extended memory. This enables applications to run in 80286 protected mode and to make use of more than 640 Kbyte of memory. Although transparent to the user, the programs still use real-mode DOS and BIOS for I/O services.

There are therefore the following three categories of memory: conventional; extended; and expanded. Memory maps differ according to what the different parts of the memory do.

When the Intel 80386 was introduced, it became possible to write DOS multitaskers. As previously mentioned, the 80386 included a virtual 8086 mode, a true 32 bit programming model, and a paged Memory Management Unit (MMU). The paging capabilities allowed the DOS multitaskers to emulate expanded memory support using extended memory. The 80386 therefore provided a platform for concurrent execution of multiple DOS applications using more than 640 Kbytes of memory. Examples of DOS multitaskers are Quarterdeck Desqview and Windows 386 (Windows 3.0). These systems are based on the non-re-entrant real-mode DOS and BIOS. As they rely on switching between the real and protected modes, many complications can arise in this multitasking environment.

Features

Advantages	Disadvantages
De facto standard	Single user only
Wide range of packages available	No multitasking
Small memory requirement	Only 640 Kbytes
Graphical user interface available (Windows)	Linearly addressable

Technical Specification

Hardware requirements
Processor   8086, 80286, 80386, 80486
At least one floppy-disk drive

Operating system
    Mode   real mode
    Memory requirement   version 3.2, approx 50 Kbytes;
                                            version 4.01, approx 70 Kbytes
    Address space   640 Kbytes, linearly accessible
    Number of users   1
    Number of active programs   1
    Interprocess communication   no

4.6.3 OS/2

The OS/2 was designed to support a single interactive user with multitasking on
a PC. One of the main reasons for multitasking was mentioned above—increased
utilization of the computer as a resource. However, another demand-driven
reason is that application packages have become much more complex and
inter-related. The user may wish to use a database, a word processor, a
spreadsheet, and a package to produce, for example, slides. Without multitasking, the user must open, use, and then close down each application, the
relevant data then being imported to each application. Any changes that must be
made require the user to repeat the whole procedure. In multitasking, each
application can be left open and is more easily transferred to other packages. A
user can therefore work with several programs at once.

The main objectives when first developing the OS/2 were the following: to
break the 640 Kbyte physical memory barrier and support up to 16 Mbytes of
physical memory; to use virtual memory to extend the physical memory
resources of the system; to provide a protected multitasking environment; to
provide an extendible, flexible system application program interface (API)
architecture; to provide a GUI; and to support the DOS application binary
compatibility to encourage migration from DOS to OS/2.

The OS/2 versions 1.X (1.1, 1.2, and 1.3) are the 16 bit versions of the OS/2
system. The OS/2 2.0 is the 32 bit version of the OS/2. The objectives for the
OS/2 2.0 were the following: to exploit the features of the 80386 and 80486
microprocessors; to provide a demand-paged system with a 32 bit programming
model, that was portable to other 32 bit processor architectures; to multitask
DOS applications in a protected environment; to provide 16 bit OS/2 application compatibility; and to provide Windows 3.0 application compatibility in a
protected environment.

The operating system has two modes, real and protected. In real mode, the
OS/2 is compatible with the DOS, as long as the application uses only standard
DOS commands. Complex applications are unlikely to meet this criteria. In real
mode, the user only makes partial use of the possibilities offered by an 80286/
80386 processor. In the protected mode, multitasking is implemented with the
aid of time sharing. This allows the concurrent processing of up to 16 application
programs and the parallel processing of operating system functions. Data
exchange between programs is possible (interprocess communication is via
queues). The processing sequence of programs can be changed by setting

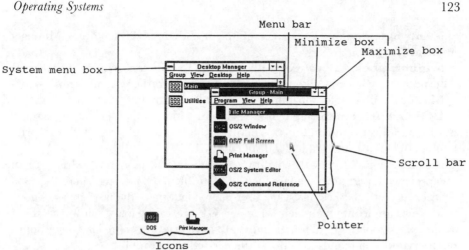

Figure 4.32 *Presentation manager screen*

priorities. If a waiting program becomes executable because of its priority, it can respond to an external event within 30 ms.

4.6.3.1 System Details. The OS/2 system consists of the kernel, device drivers, dynamic link libraries, and application programs. The heart of the OS/2 system is the kernel. The kernel contains the control program that runs with supervisor privileges. As in the DOS, the kernel uses device drivers to access the hardware resources. The most critical parts of the system reside in the kernel. These include multitasking, memory management, interprocess communication, DOS compatibility, and I/O control. While the kernel also contains some of the system's APIs, others are located in dynamic link libraries. These are shared libraries, that can be used to extend the functionality of the system. The locations of the APIs are transparent to the applications.

4.6.3.2. User Interface—Presentation Manager. The Presentation Manager (PM) is the GUI of the OS/2. It provides a medium for applications to share the video display in a windowed environment (Figure 4.32). Presentation Manager programs have a common user interface composed of windows, scroll bars, dialogue boxes, pull-down menus, and other desktop controls that are accessed through the keyboard, mouse or other user input devices. This user interface is easier to learn and more consistent than traditional line oriented user interfaces.

Pointer: a mouse will give you an arrow-shaped pointer that shows your position on the screen.

Scroll bar: a scroll bar may appear in the window or dialogue box that contains more information than can fit into the work area.

System menu box: used to display the system menu, which is a list of commands common to all groups and applications that run in a window. It can be used to move and alter the size of windows, quit applications, and close groups.

Menu bar: displays the names of the menus in a Presentation Manager application or group.

Minimize box: can be used with a mouse to reduce a window to an icon. Alternatively, you can use the minimize command on the system menu.

Maximize box: as for the minimize box, but used to increase the window size.

DOS and print manager icons: represent windows that have been reduced by the minimize command/box. When the OS/2 starts up the DOS and print manager icons are automatically displayed.

A common problem with operating system interfaces is that infrequent or inexperienced users are often faced with the problem of knowing where they are. Graphical user interfaces assist the user in 'navigation'—the user always knows where they are within the system—further, GUIs are easy to use. However, this is at the expense of depth of functionality. Experienced users prefer to issue short, often cryptic, commands to perform the desired tasks rapidly.

4.6.3.3 Process Management. OS/2 uses a pre-emptive scheduler with round robin scheduling and multiple-level priority. The four main classes of priority are time critical, foreground jobs, regular class, and idle class.

4.6.3.4 Memory Management. Up to 16 Mbytes of main memory are directly addressable. If the amount of memory available is less than 16 Mbytes, the OS/2 stores part of the programs on the secondary memory. Disk operating system applications can be used without any problems. The memory management model presented by the 16 bit versions of the OS/2 is a model that uses the 80286 virtual memory capabilities. The system runs in 80286 protected mode, and can therefore make use of up to 16 Mbytes of physical memory. This is a signficant breakthrough of the 640 Kbyte barrier associated with the DOS/8088 systems. The segmented memory model of the 80286, with a maximum segment size of 64 Kbyte, must still be considered. The system uses the 80286 memory-protection features to protect system memory from user memory, and to protect individual processes from each other. The sharing of memory amongst processes is also supported by the OS/2 memory manager. This is a powerful form of interprocess communication, and is important to the architecture of shared libraries and subsystems.

4.6.3.5 Compatibility with the DOS. One of the most critical features is the capability to run DOS programs. To capture the unprotected DOS environment in a system such as the OS/2 that provides traditional resource management is a difficult problem. This problem is made worse by the fact that the 80286 system does not have the capability of running 8088 applications in protected mode. The goal for the 80286 OS/2 systems is to allow a single DOS application to run in the foreground while OS/2 applications run in the background. The DOS application does not benefit from the multitasking facilities; when it is moved to the background, the execution is frozen. The DOS environment exists in a special session called the DOS session. A technique called *mode switching* is used to support the concurrent execution of a DOS application in the foreground and OS/2 applications in the background. This involves switching from protected to real mode and back. The switch from protected to real mode is accomplished with the assistance of help circuits on 80286 machines. Because of the mode switching, certain critical parts of the OS/2 kernel and device drivers must be accessible in both real and protected mode. Code that can run in both real and protected mode is called bimodal code. To be addressable in both modes, the bimodal code must reside in the physical memory below 640 Kbytes.

4.6.4 OS/2 2.X

At the time of writing, the OS/2 2.0 was the only version of the 32 bit OS/2 system. This system is targeted at the 80386/80486 processors. It uses the paging features of the 80386 to provide a demand-paged virtual-memory environment that supports a new 32 bit portable programming model. The OS/2 2.0 provides binary compatibility for OS/2 1.X applications. Compatibility with the DOS is enhanced by using the 80386 virtual 8086 mode to enable multiple DOS sessions to run concurrently and in the background. Generally, the system contains the same as the 1.X systems, but it is scaled up to 32 bit, and has an architecture designed for portability to other processor platforms. The user shell has also been improved, providing a PM based installation program and the ability to perform installations across a LAN. Also, a new user shell called the workplace shell provides an object-oriented environment. The OS/2 2.0 runs all OS/2 1.X applications without change. An architecture has been designed in which 16 and 32 bit modules can coexist.

1.6.1.1 Compatibility with the DOS. The OS/2 2.0 uses the 80386 virtual 8086 mode, and paging to provide multiple DOS-compatible environments. The entire DOS environment is encapsulated in a virtual DOS machine (VDM), giving the system better protection than offered by the OS/2 1.X. In the OS/2 2.0, a DOS application can only hang its own session, instead of the entire system. The errant DOS session can then be terminated from the desktop manager. The DOS applications can be run full screen, windowed or iconized in the background. The system gives these applications about 620 Kbytes in which to run, more than that usually provided by the DOS. Expanded memory specification and expanded memory support (XMS) are provided. Specific versions of the DOS can be booted into different VDMs. The environment can be tailored to emulate any DOS environment.

Features

Advantages	Disadvantages
Multitasking capability	Single user only
Real and protected mode	Large memory requirement
16 Mbyte memory capacity	Few applications available
Good networking capability via	MS-DOS applications may give
the LAN manager	some problem
Graphical user interface available	
(PM)	

Technical Specifications

Hardware requirements
    Processor   80286, 80386, 80486
    Main memory requirement   about 4 Mbytes
    One hard-disk drive   at least 40 Mbytes
    VGA graphics controller
Operating system
    Modes   real or protected mode
    Address space   16 Mbytes linear; 1 Gbyte virtual
        Response time   < 30 ms
        Number of users   1
        Number of concurrently
        executable programs   1
        Interprocess communication   via queues
        Maximum program length   16 Mbytes

4.7 OPERATING SYSTEMS—CATEGORIES AND EVALUATION

Operating systems can be classified according to the type of data processing they are best suited for, *batch* or *demand* (Figure 4.33). In batch data processing, the user submits a job or jobs for processing, with the results being supplied to the user at a later date. In demand-driven data processing, the requirement is that the data are processed 'on demand'. Complex operating systems allow batch jobs to be run in the background while the demand jobs are in the foreground. The most common type of demand processing is *interactive*, where users interact directly with the computer via a terminal. In *real-time* processing, there must be a guaranteed response time appropriate to the application as the results are used to control devices such as reactor vessels, *etc.*

On-line: an electrical connection between the data and the computer. The computer can access the data.

Batch processing: a predefined processing period.

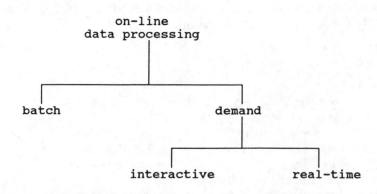

Figure 4.33 *Data processing environment*

Table 4.6 *Operating system summary*

System	Multiuser capability	Multitasking capability	Real time	Maximum memory management
DOS	no	no	no	640 Kbytes
OS/2	no	yes	partial	1 Gbyte
UNIX	yes	yes	no	16 Mbytes

Demand processing: data processing 'on demand'. The data are processed as soon as possible.

Interactive: interactive dialogue between the computer and the users. The response time is appropriate to the user.

Real-time: the computer must guarantee to respond to the data input within predefined time constraints.

Evaluation methods vary. Simulation techniques use a model of the computer system with a random number generator that provides figures for the CPU time, job arrivals, *etc*. Trace tapes are created by monitoring a system and recording actual events. Trace tapes can then be used to drive simulators. The ultimate test though is on-site implementation and evaluation.

Responding to certain situations within defined delays can be an important part of automation. Many tasks can only be automated if the operating system is able to prioritize data, *i.e.* critical conditions arising, which must be responded to in a guaranteed maximum time.

For a system to respond under real-time conditions, the parts of the program that the response requires must be resident in the main memory. The time taken to swap a current program out onto disk and the new program into the IAS is too long. Further, the CPU scheduling algorithm must allow high priority tasks to be completed 'on demand'.

CHAPTER 5

Input and Output Interfacing—the Outside World

The word *interface* has a wide range of meanings and may be applied to hardware, software, individual chips, large systems, *etc.* In our context we are taking it to mean the hardware and software necessary to form a buffered link between the computer's internal buses and the outside world. An interface is a connection or boundary between a controlling device (a computer) and a connected device (I/O device). The ANSI/IEEE Std 729, 1983, defines an interface as: 'A shared boundary. An interface may be a hardware component to link two devices or it may be a portion of storage accessed by two or more computer programs'. To interface is defined as: 'To interact or communicate with another system component'.

More specifically, an interface is the definition of the logical, electrical, and physical properties of the boundary. We have already met the I/O interface in Chapter 2 (Figure 5.1). There is the requirement for international standards that define electrical and mechanical specifications thereby allowing equipment from

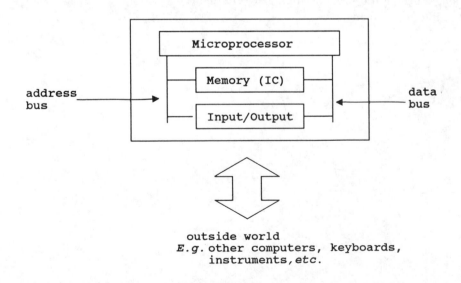

Figure 5.1 *Microcomputer I/O*

different manufacturers to be connected together. The EIA has produced several standards that will be considered shortly.

5.1 SIMPLE COMPUTER INPUT AND OUTPUT

Recall our memory map from Chapters 2 and 3 (Figure 5.2). We saw how, not only our program was stored in the memory, but also that data could be stored in uniquely addressable locations. For simplicity, the control bus lines are not shown in Figure 5.2. The computer struture can be simplified by showing how the memory (RAM and ROM) and the CPU are all connected to the three buses (Figure 5.3): the Address Bus (AB) has 8 lines, the DataBus (DB), 8 lines, and the control bus, n lines. From now on we will refer to Address Bus, Data Bus, and Control Bus as A, D, and C respectively.

It is possible to have an addressable data location that is connected to the outside world via wires. This can be achieved by using an I/O chip that is connected directly to the system address, data, and control lines—in exactly the same way as ordinary RAM and ROM memory chips (Figure 5.4). The I/O chip has uniquely addressable memory locations—exactly as we saw with the normal RAM and ROM memory. To be more precise, we can therefore address the contents field of the I/O chip and read/write data accordingly. However, as we

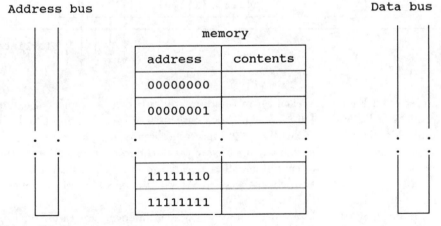

Figure 5.2 *Memory*

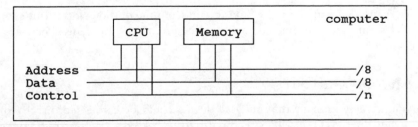

Figure 5.3 *Address, data, and control lines*

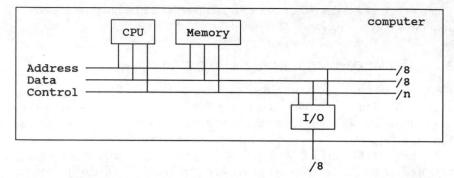

Figure 5.4 *I/O unit*

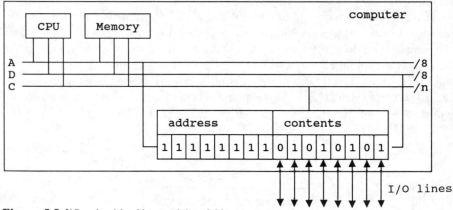

Figure 5.5 *I/O unit with address and data fields*

can see from Figure 5.5, each bit in the data contents field is connected to the outside world. For simplicity, we will allocate memory address location 11111111 to this data field (Figure 5.6). We could have chosen any address.

The data contents field that is connected to the outside world is therefore in the *memory space* of the computer. This is called *memory mapped I/O*. The instructions in the instruction set can be used to access I/O data. It is possible to have several such addressable locations, as we will see shortly.

Alternatively, the address space may be kept isolated or separate from the normal memory space. This is called *isolated I/O*. The same bus lines are used but additional lines are provided, and additional instructions are needed in the instruction set.

We can now read and write data to and from the outside world! Immediately we are connected to the outside world we begin to have problems. These are now considered.

5.2 INTERFACE UNITS

It is extremely unlikely that our I/O device could be connected directly to the I/O data lines. There will be compatibility problems, such as mismatches in

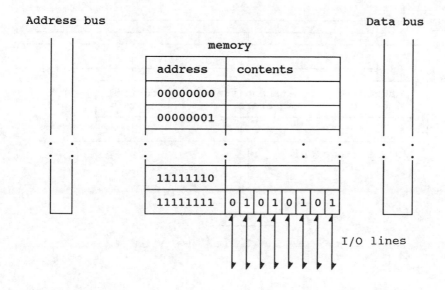

Figure 5.6 *I/O memory address*

speed between the CPU and the I/O device, that will be considered in detail shortly. Further, there are very many different types of I/O device, all with different characteristics, that we may wish to interconnect. To simplify matters we therefore need an *interface unit* that enhances the functionality of the simple I/O chip (Figure 5.7). This relieves the CPU of direct control of the I/O device and allows the CPU to issue standard commands. Interface units vary in cost and complexity. Typically, they have a separate processor dedicated to the I/O and are sometimes referred to as I/O processors (IOPs). As such they have a simple instruction set and architectural features optimized for I/O operations. At a minimum, they have simple data registers for storage, but may even have RAM. Figure 5.8 shows in greater detail the functional architecture we may have.

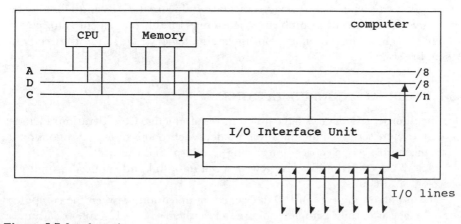

Figure 5.7 *Interface unit*

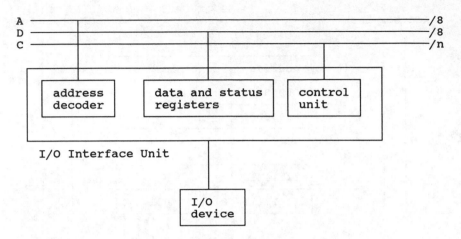

Figure 5.8 *Interface unit architecture*

The address comparator (AC) or address decoder allows the CPU to access individual I/O interface units (IUs). The desired address is placed on the AB but only one device can directly decode this address.

The CU is responsible for the detailed movement of data between the computer bus and the external device.

Typically, two data registers will be used to hold data that are passed to and from the I/O device. The registers hold data until they can be used. Associated with each data register will be a control register and a status register. More complex I/O interface units can support several I/O devices as shown in Figure 5.9.

5.3 MULTIPLE-BUS ARCHITECTURE

With a large number of I/O devices, the data traffic on the ABs and DBs may be excessive. To address this problem, one architectural arrangement is to have a separate memory bus specifically for the main memory that connects directly to the CPU (Figure 5.10).

5.4 INTERFACE CONSIDERATIONS

As mentioned, I/O devices have different requirements. Considerations include the bit or byte serial, communications mode, synchronous or asynchronous data transmission, speed, electrical interfacing standard, and protocol.

Some I/O devices need a single wire serial data link, others need a multiple wire or parallel link.

Depending on the type of I/O device, communication between the computer may be one way only (simplex), alternately either way (half duplex) or simultaneously both ways (full duplex).

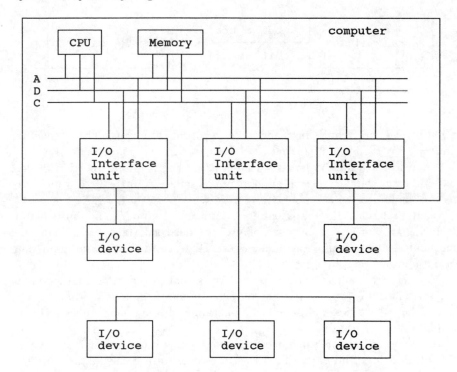

Figure 5.9 *Complex I/O interface units*

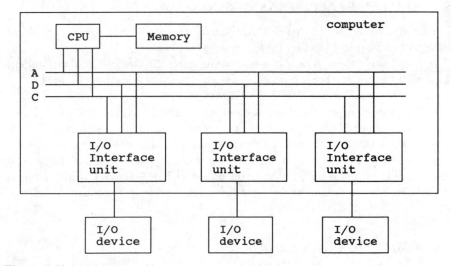

Figure 5.10 *Multiple bus architecture*

The I/O device may send a continuous data stream (synchronous transmission) or occasional bytes of data (asynchronous transmission).

Typically, I/O devices work slowly compared with the CPU. Does the CPU idle while waiting for data input, and if so, what about CPU utilization? If the

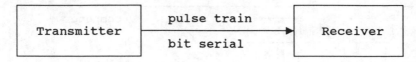

Figure 5.11 *Serial transmission*

CPU is busy, how can it be made to react to critical events? These issues are addressed by *polling* and *interrupts*.

What electrical standards exist for interfacing? For example, some I/O devices have different voltage representations of binary data, *e.g.*, $+/-30$ V, which may not be compatible with our system.

What mechanical standards exist for interfacing? In order to minimize manufacturing cost and to allow ease of use, we need standards that define the mechanical connections we are going to use. Do we need a square three-pin plug or a round two-pin plug?

Protocol is the set of conventions or rules that govern the interactions of processes or applications within a computer system or network. Alternatively, this is the set of rules that govern the operation of functional units to achieve communications (ISO).

The above criteria are sometimes grouped together into three main categories, logical, electrical, and physical (LEP). Let us now address each of these in turn.

5.5 SERIAL/PARALLEL TRANSMISSION

The cabling between two pieces of equipment may allow data to be transmitted either bit serial or bit parallel (byte serial).

In serial mode there is only one wire connecting the two pieces of equipment for data transmission. Each bit of the pulse train must be sent in turn, one bit at a time, *i.e.*, bit serial as shown in Figure 5.11.

Within a small room, the distances and hence lengths of cable used to connect pieces of equipment together are short. It is possible to use a separate piece of wire for each data bit. Therefore there are multiple wires connecting equipment together, *i.e.*, for parallel data transfer. If we have, say, eight wires, we can send 8 bits at a time, which is bit parallel mode (Figure 5.12). Do note that this mode is also called byte serial. This is faster but more expensive; minimum delay is ensured, but at maximum cost.

5.6 COMMUNICATION MODES

There are different modes of communication. With a very talkative person the conversation is one way only! In normal conversation there will be two-way communication with each speaker taking turns to talk. In communication engineering there are in fact the following three types of communication: simplex, in which data are transmitted in one direction only, *e.g.*, a data-logging device); half duplex, in which data may be exchanged between two devices, but

byte serial
bit parallel

Transmitter	bit 1	Receiver
	bit 2	
	bit 3	
	bit 4	
	bit 5	
	bit 6	
	bit 7	
	bit 8	

Figure 5.12 *Parallel transmission*

only one device can transmit at a time (*e.g.*, programmable instrumentation); and duplex, which allows data exchange between two devices in both directions simultaneously.

5.7 SYNCHRONIZATION

We saw the pulse train in Chapter 1 (Figure 5.13). Character transmission is started with the start bit, which is always 0, and ends with a stop bit(s), which is(are) always 1. The transmitting device has a clock that generates clock pulses at a regular frequency. The clock pulses define the duration of each bit pulse, the *bit period* (t) (Figure 5.14). The number of clock pulses per second is called the

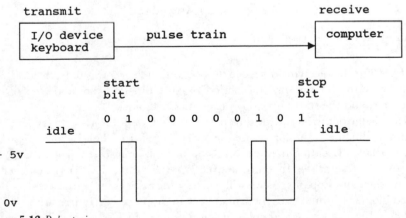

Figure 5.13 *Pulse train*

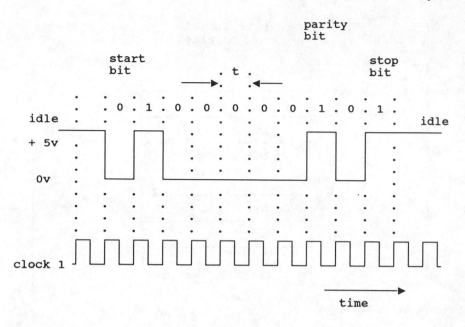

Figure 5.14 *Pulse train frequency*

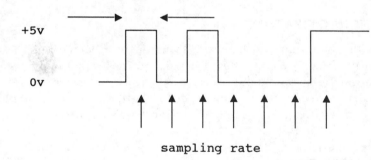

Figure 5.15 *The correct sampling rate*

baud rate. For the receiving device to decode and interpret this pattern correctly it must know the bit rate or time period of each bit, the start and stop of each character, and the start and stop of each message block.

Knowledge of the bit rate or time period of each bit allows the receiving device to sample the line at the correct frequency (Figure 5.15). This is called *bit* or *clock synchronization*. Locating the start and stop of each character is known as *byte* or *character synchronization*. In our example from Chapter 1, this is achieved by having start and stop bits. As we will see, in some devices large quantities of data, called blocks or frames, must be sent at very high speed. It is important to be able to define the beginning and end of each message block or frame. This is called *block* or *frame synchronization*. In effect, the receiving device must be informed of the start and end of the data and also how fast the line should be sampled. The two principle transmission methods that address these issues are *asynchronous transmission* and *synchronous transmission*.

5.8 ASYNCHRONOUS TRANSMISSION

This method is used when there are long idle periods between characters that are being sent at random intervals, *e.g.*, from a keyboard. The transmission line will be idle for extended periods of time and the receiver must be resynchronized at the start of each new character. This is achieved by enveloping the data between start and stop bits. By convention, the idle line is held high and the first 1 to 0 transition indicates to the receiving device the start of a new character. The polarity of the start and stop bits differ thereby guaranteeing a transition between characters. As a single wire connects the transmitter and the receiver (simplex, half duplex), it is not possible to transmit both the data and the clock signal. Instead, the transmitter and receiving clocks work independently, *i.e.*, the receiver does not know the correct frequency of the transmitting clock and so the exact bit period is unknown to the receiver (Figure 5.16). In this case, the clocks (clock 1 and clock 2) will not be synchronized. For example, with the two clocks running at the same frequency, bit-cell periods may not coincide or indeed they may be working at different frequencies (Figure 5.17). The result of this may be that the data are sampled at the bit-cell boundaries. This is the worst place to sample digital information due to the changing nature of the signal at this point (Figures 5.18 and 5.19). The problem is solved by running the receiving device clock at a higher frequency than the transmitted bit-rate frequency. This allows the receiver to identify the approximate mid-point of each transmitted bit. As we saw earlier, character transmission is started with the start bit, which is always a 0. This alerts the receiver that a character is being received. The stop bit is

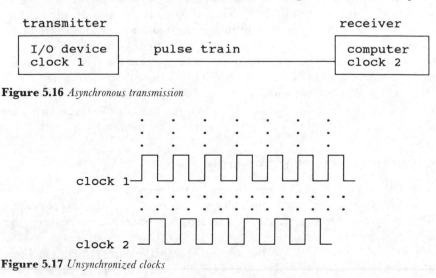

Figure 5.16 *Asynchronous transmission*

Figure 5.17 *Unsynchronized clocks*

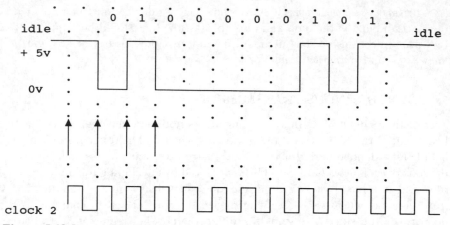

Figure 5.18 *Incorrect sampling points*

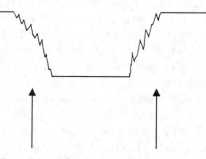

Figure 5.19 *Worst case sampling points*

always 1. What actually happens is that the receiver clock runs at 16 times the frequency of the transmitter clock. This allows the receiver to identify the approximate mid-point of each transmitted bit. The receiver samples the logic level of the line at every receiver clock pulse until it finds eight successive 0s. As 1 cycle of the transmitter clock is equivalent to 16 cycles of the receiver clock, the eight 0s identify the centre of the start bit. The receiver then successively counts 16 receiver clock pulses and samples once at the count of each group of 16 pulses. This continues until the stop bit (Figure 5.20) is reached. In effect, the pulse train is sampled at the mid-point of the bit-cell period as shown in Figure 5.21.

5.9 SYNCHRONOUS TRANSMISSION

Asynchronous transmission is normally used when there are long, indeterminate delays between data transmissions between characters. In many applications, however, there is the requirement to transfer large data volumes at very high speed, *e.g.*, between computers or fast I/O devices. Using the start and stop bits would add a 20% overhead. Further, use of the scaled clock to achieve synchronization gives a maximum reliable rate of 19 200 bps. A more efficient scheme is

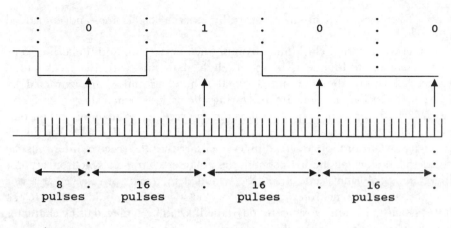

receiver
clock pulses

Figure 5.20 *Asynchronous data sampling*

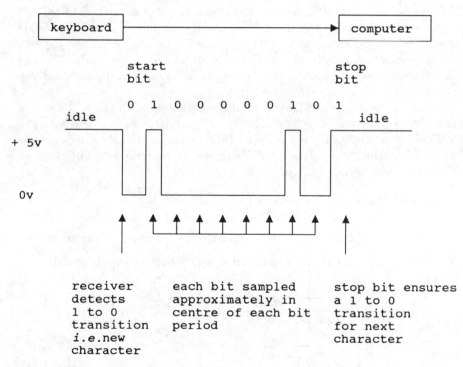

Figure 5.21 *The full pulse train with sampling*

therefore used that transfers complete data blocks as single units with no delays between each character—synchronous transmission. The data are enclosed by start of frame (SOF) and end of frame (EOF) characters. With this type of data transmission both transmitter and receiver clocks are synchronized. The two ways of achieving synchronization are by external clock line, and embedded clock.

Use of the external clock line obviously necessitates an additional line. The more common technique is to use a single data line with the clock pulses embedded within the transmitted signal. The clock pulses are extracted by means of a clock-extraction circuit. During idle periods, synchronization (SYN) characters are sent continuously, allowing the receiver to maintain synchronization. The requirements are therefore the following: all frames should be preceded by one or more reserved bytes to ensure that the receiver interprets the correct character for the byte boundaries (byte or character synchronization); the contents of each frame should be bounded by a pair of reserved bytes or characters (frame synchronization); and a clocking system should operate (clock or bit synchronization). A synchronous data link can be achieved using character (byte) oriented or bit oriented schemes.

5.9.1 Character-oriented Synchronization

With this scheme each data frame consists of a variable number of characters (each 7 or 8 bits) that are transmitted as a contiguous pulse train with no delays between them. The receiving device must therefore be able to achieve clock or bit synchronization using a suitable clock encoding scheme, detect the start and stop of each character (character synchronization), and detect the start and stop of each data frame (frame synchronization).

To achieve character synchronization, the transmitting device sends two or more special SYN characters immediately before data frame transmission. The SYN character (00010110) is a reserved ISO character. The receiver scans the input line looking for the known SYN character thereby achieving byte or character synchronization (Figure 5.22).

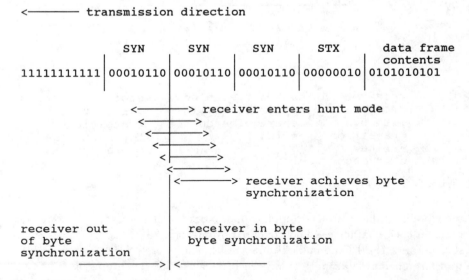

Figure 5.22 *Byte synchronization*

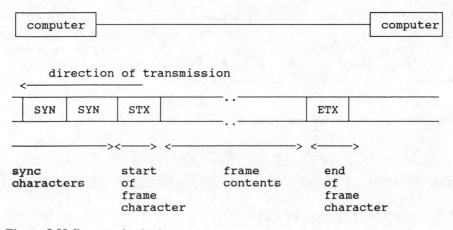

Figure 5.23 *Frame synchronization*

Similarly, characters from the ISO character set define frame limits, *i.e.*, the start of text (STX) and the end of text (ETX). After character synchronization and receipt of the STX, each character is compared with the ETX. If it is not the ETX, the character is a data item and is stored accordingly (Figure 5.23).

It is important that the synchronization process is independent of the contents of the data frame, *i.e.*, *data transparent*. In some applications the data frames may be binary files rather than printable characters. Unfortunately, the binary sequence for our ETX character may be in this binary file! This would cause incorrect termination of the data frame. To achieve data transparency, a pair of characters is used both to signal the start of a data frame and the end. Furthermore, should the transmitter detect, for example, a data link escape (DLE) character in the binary file, it would insert a second DLE character into the data stream. This is called character or byte stuffing (Figure 5.24). The receiver can detect the EOF (End of File) by the unique DLE–ETX sequence. Furthermore, whenever it receives two DLE characters the second is discarded. When data are

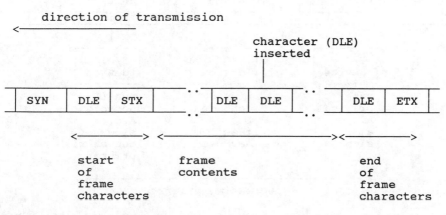

Figure 5.24 *Character stuffing*

transmitted over a distance it has been found that groups or strings of errors occur. These are error bursts. Parity does not provide protection against error bursts; however, polynomial codes can be used when data frames are transmitted. As we saw in Chapter 1, the check digits are called the Frame Check Sequence (FCS) or the Cyclic Redundancy Check (CRC).

5.9.2 Bit-oriented Synchronization

With this type of scheme, the transmitted data frames are not restricted to sending multiples of 8 or 7 bits, any arbitrary number of bits may be sent. These are less common, however, and will not be further considered here.

5.10 INTERRUPTS AND POLLING

When your computer is multitasking, not only is it responding to the keyboard, it is simultaneously servicing I/O devices. With few exceptions, external devices (*e.g.*, slow dot matrix printers and keyboards) run at slower speeds than computers. Indeed there are a wide range of devices that work at different speeds and with different priorities (Figure 5.25). Input/output devices can be classified according to their relative priority.

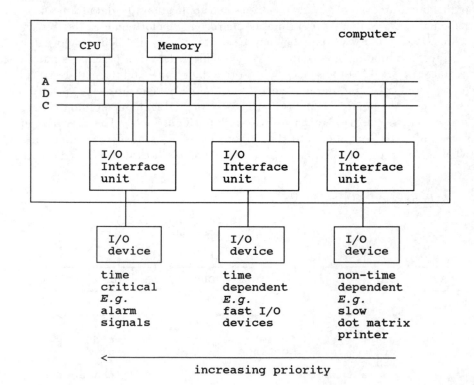

Figure 5.25 *Prioritized devices*

Several questions were raised earlier. Does the CPU idle while waiting for data input? If so, does CPU utilization decrease as a consequence? If the CPU is busy doing something else, how can it be alerted to critical events? With multiple devices, how can each device be serviced according to its needs and priority? These issues are addressed by *polling* and *interrupts*.

5.10.1 Polling

Polling is the regular interrogation of devices to determine whether they need attention. Each interface unit has a *flag* that can be set when servicing is required. A program regularly interrogates or *polls* each device in turn checking the state of the flags (Figure 5.26). Different algorithms exist to take priority into account (Figure 5.27). Note that the positional priority algorithm may lead to device starvation. Whilst this scheme is simple, it is slow and wasteful as a lot of processor time is spent interrogating each device.

5.10.2 Interrupts

The processor must be able to interact efficiently with external devices. Recall the pin-out diagram for the microprocessor we met in Chapter 2 (Figure 5.28). In particular we identified two pins: the interrupt request (IRQ) and the non-maskable interrupt (NMI).

These pins allow external devices to *interrupt* the microprocessor. An interrupt

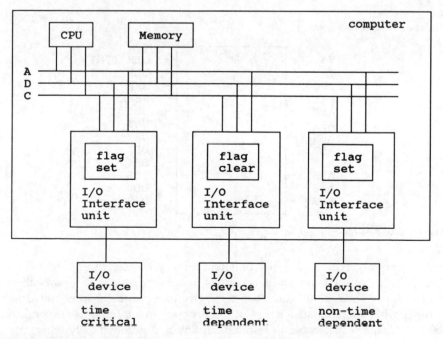

Figure 5.26 *Software polling flags*

Equal priority   Positional priority

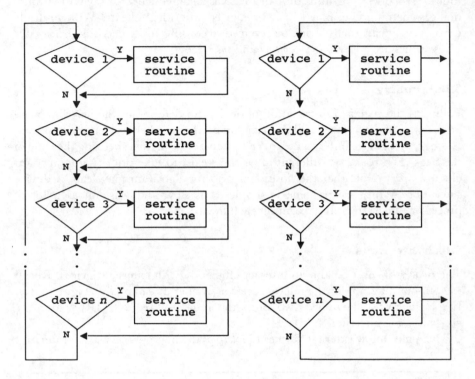

Figure 5.27 *Priority schemes*

V+	CLK (IN)
V-	CLK (OUT)
IRQ	R/W
NMI	RESET
AB0	DB0
AB1	DB1
AB2	DB2
AB3	DB3
AB4	DB4
AB5	DB5
AB6	DB6
AB7	DB7

Figure 5.28 *Microprocessor pin out diagram*

is an externally generated signal that temporarily suspends the program currently being executed and transfers control to another (small) program called a *routine* or a *subroutine* that is designed to service that particular interrupt. Interrupts provide an automatic change in the program as a result of some condition arising. Hence it is possible to respond rapidly and automatically whenever a significant event occurs. These devices can inform the processor that they need

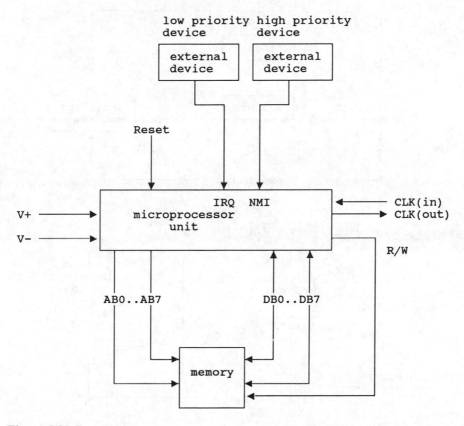

Figure 5.29 *External interrupts*

attention by using these two interrupt lines. The IRQ is an interrupt request that may or may not be given attention, *i.e.*, a low priority request, and the NMI is an interrupt request that cannot be ignored (Figure 5.29).

5.10.2.1 Interrupt Cycle. When an interrupt occurs the program currently being executed is halted and control is transferred to a subroutine that is designed to service that particular interrupt (Figure 5.30). On completion of that sub-routine, control is returned to the program that was interrupted (Figure 5.31). The interrupt allows a device or user to deflect the processor into executing another program without having to wait for the current program to be completed. In effect, this is polling at the machine level. The exact details vary between processors; however, a typical sequence of events is as follows: the processor's interrupt enable flag is set in the processor status register (condition code register); the current instruction is finished; the register's contents are saved, *i.e.*, freeze status; the interrupt source is determined; the interrupt service routine is executed; the interrupt flag is reset; and the original status is restored.

5.10.2.2 Interrupt-based Software Polling. With the IRQ pin, the microprocessor can attend to other tasks until the IRQ pin is activated. Even though there may only

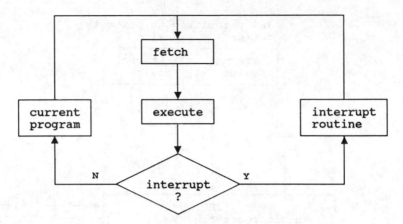

Figure 5.30 *The interrupted fetch execute cycle*

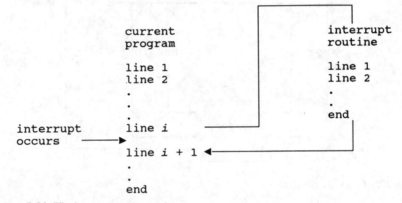

Figure 5.31 *The interrupt service routine*

be one IRQ line it is possible for many devices to use this single line. Software will then interrogate each device (Figure 5.32). For simplicity the memory is not included in Figure 5.32.

5.10.2.3 Vectored Interrupt. A faster but more expensive approach is to use *vectors*. Each interface unit has a register with a binary value that identifies the start address of its own service routine in memory. This address is called the vector address or vector. On receipt of the interrupt, the processor issues an acknowledge on the ACKnowledge line. This acknowledge is passed through each device in turn until it reaches the devices that signalled the interrupt (Figure 5.33). This is called *daisy chaining*.

5.10.2.4 Interrupt Masking. A more dynamic and flexible system is to have a *mask register* and *priority logic controller* (Figure 5.34). The bit pattern in the mask register is under software control. The bit pattern chosen can selectively *mask out* devices. Recall the logical function of an AND gate—a 1 in the input will always give an output of 0.

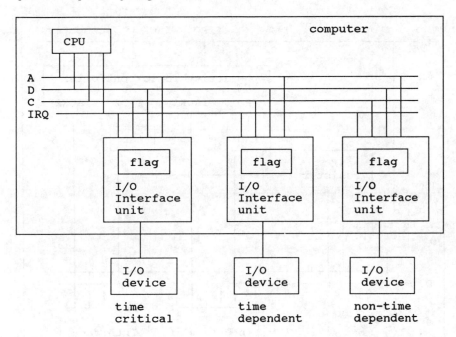

Figure 5.32 *Interrupt based polling*

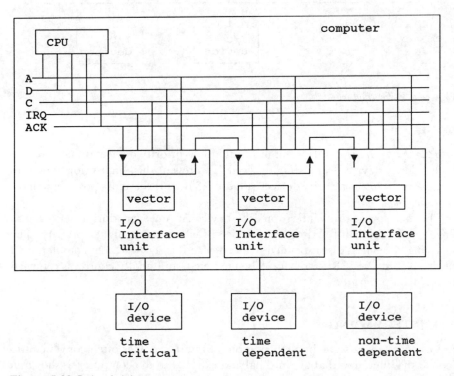

Figure 5.33 *Daisy chaining*

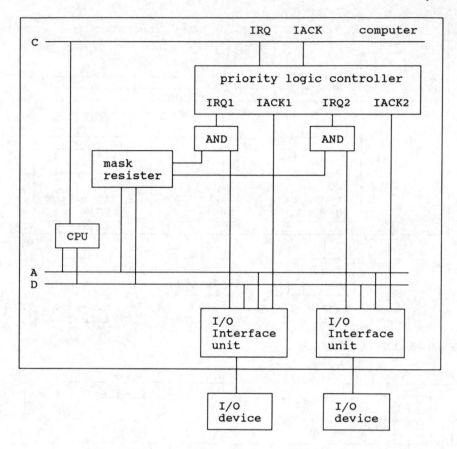

Figure 5.34 *Interrupt masking*

5.10.2.5 Nested Interrupts. It is important that low-priority devices do not prevent high-priority devices from being serviced. To prevent this, high-priority devices can interrupt lower priority devices (Figure 5.35). It is therefore possible to have *nested interrupts*.

The main program is interrupted by device A; however, during servicing of device A an interrupt arrives from device B, which has a higher priority. The servicing of device A is suspended whilst device B is serviced. On completion of device B ISR, the ISR for device A is completed, and finally control is returned to the main program.

5.11 INTERFACING

Let us now look at some implementations of interfaces. It is often convenient to look at the interfaces that are internal and external to the computer as shown in Figure 5.36.

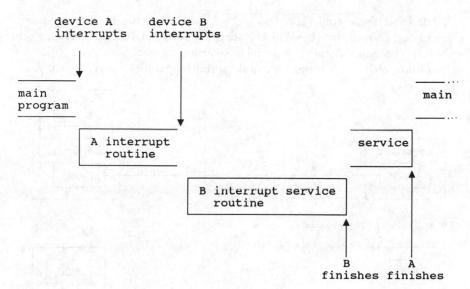

Figure 5.35 *Nested interrupts*

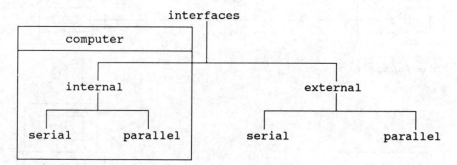

Figure 5.36 *Interfaces*

5.11.1 Internal Interfaces—The Serial Link

There are special IC chips that are available for the different types of serial and parallel interface as shown in Figure 5.37.

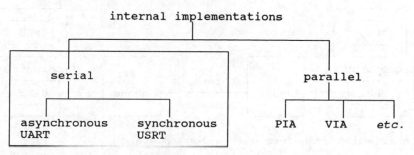

Figure 5.37 *Internal interface implementations*

5.11.1.1 Asynchronous Serial Transmission—the UART. The universal asynchronous receiver and transmitter (UART) is a specially designed IC that performs the following functions: parallel to serial conversion of each character prior to transmission on the bit serial data link; serial to parallel conversion of each

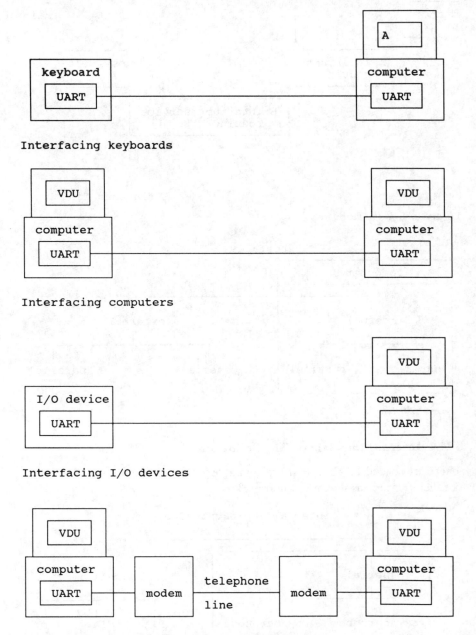

Interfacing keyboards

Interfacing computers

Interfacing I/O devices

Interfacing computers over a long distance

Figure 5.38 *UART applications*

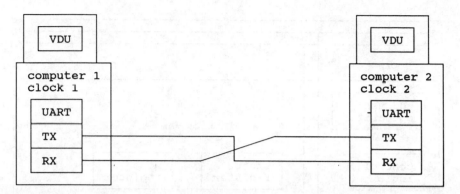

Figure 5.39 *UART transmit and receive units*

received bit and conversion to character format; assists the receiver in achieving bit or character synchronization; and error checking.

These ICs can be used in a variety of applications as shown in Figure 5.38. Each UART has a transmit (TX) and receive (RX) part as shown in Figure 5.39. Communication is character based, typically using the ASCII code with the least significant bit being transferred first.

The UART is a programmable device that allows the user to specify certain operating characteristics. The functional units are illustrated in Figure 5.40. There are also on-chip modem controls. Modems will be discussed shortly.

The mode or control register has to be loaded with the required bit pattern defining the operating characteristics (Figure 5.41). This is sometimes called initialization. Often, initialization has already been done for you. The pattern choices include the following: the number of bits per character; odd, even, or no parity; the number of stop bits; the transmit bit rate; and the receive bit rate. The controlling device is able to determine the status of the UART by reading the contents of the status register and testing each bit accordingly (Figure 5.42). These are often called flag bits.

To transmit a new character the controlling device must first read the status byte to determine the state of the transmit buffer (TXB F/E). If the buffer is empty, the previous character has been transmitted and the device is able to send another one. The character to be sent is put into the transmit buffer by the control logic and the parity is set accordingly. The character is shifted bit serially into the transmit register. When a new character is placed in the TXB the TXB flag is set accordingly. Transfer of the character to the TX register resets the flag.

For receiving a character, the UART must be initialized in the same way as the transmitting UART. When the control logic detects the first 1 to 0 transition on the line, the receiver timing logic must be synchronized. This is done by the control logic presetting the contents of the bit rate counter to half that of the clock rate ratio setting. For example, if the UART has been programmed to operate with a $\times 16$ clock rate, a modulo 16 counter would be used, initially preset to 8 on the first 1 to 0 transition. As there are 16 clock cycles for each bit cell at the $\times 16$ rate, the counter will reach zero at the centre of the start bit cell

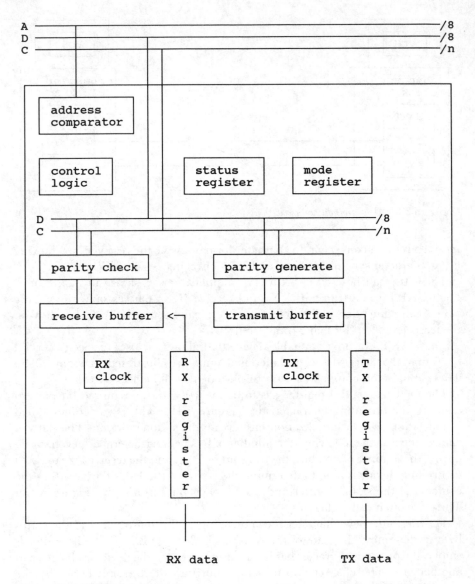

Figure 5.40 *UART functional units*

period. The bit-rate counter is then preset to 16 clock cycles and will reach zero at the centre of each bit-cell period. Every time the counter reaches zero, the control circuit is triggered to sample the current state, *i.e.*, 0 or 1, of the receive data line, and the sampled value is placed in the receive register. The higher the clock-rate ratio, the nearer the sampling point is to the centre of the bit-cell period. This process continues until the defined number of bits is loaded into the receive register. After transfer to the receive buffer, the parity is checked. If there is an error the parity error (PE) flag bit is set in the status register and the receive buffer full (RX BF) flag is also set.

mode register

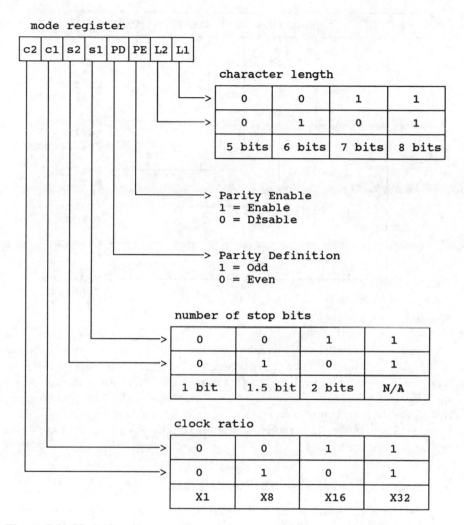

Figure 5.41 *The mode register*

The framing error (FE) flag is set if the control logic determines that there is not a valid stop bit. The overrun error (OE) is set if the controlling device has not read the previously received character from the buffer before being overwritten by the next character. Normally UARTs have a receive and a transmit section allowing full duplex data transmission. Many also have additional control lines to allow modem interfacing.

5.11.1.2 Synchronous Serial Transmission—the USRT. Special ICs are readily available for the transmission of both character- and bit-oriented schemes. The interface circuit used for character-oriented transmission is called a universal synchronous receiver and transmitter (USRT). The USRT is a programmable device that allows the user to specify certain operating characteristics. Func-

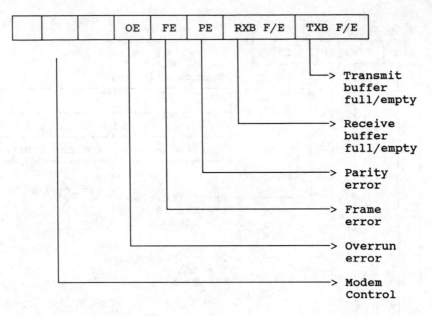

Figure 5.42 *The status register*

tionally it is very similar to the UART with the main differences lying in the use of the SYN characters. Before transmission, the SYN characters are transmitted followed by the data frame. Between data frames, the USRT automatically sends the SYN characters until the next data frame is ready. Called *interframe time-fill*, this allows the two devices to remain in synchronization. At the receiving USRT, hunt mode allows synchronization to be achieved; there is a flag bit in the status register to indicate this.

5.11.2 Internal Interfaces—The Parallel Link

A parallel interface will allow high-speed data transfer to take place and is therefore suitable for an interface to another computer or a very fast peripheral (Figure 5.43). Further, it can also be used to control actuators and monitor equipment status values. There are a wide range of chips available, including the versatile interface adaptor (VIA), the peripheral or parallel interface adaptor (PIA), and parallel input and output (PIO).

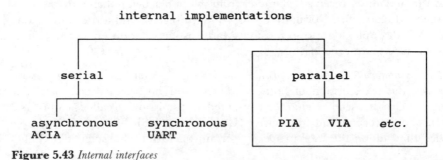

Figure 5.43 *Internal interfaces*

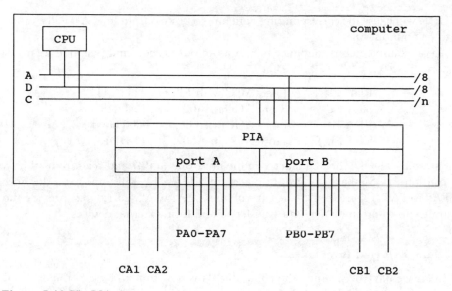

Figure 5.44 *The PIA*

5.11.2.1 Peripheral or Parallel Interface Adaptors. The peripheral or parallel interface adaptor is a memory-mapped device with two bidirectional 8 bit registers called port A and port B as illustrated in Figure 5.44. Each side of the PIA has two main registers (Figure 5.45), the port data registers [*i.e.*, port A data (PAD) and Port B data (PBD)], and the data direction register (DDR) [*i.e.*, port A data direction register (PADDR) and port B data direction register (PBDDR)].

The port data register is a temporary storage location that holds a data byte. The port A data (PAD) register has the address FF, and the port B data (PBD) register has the address FE. The individual lines are identified as PA0 to PA7 and PB0 to PB7. The DDR allows individual data lines to be configured either as input or output. The PADDR has the address FC, and the PBDDR has the address FD. Each bit in the DDR controls a corresponding data line. A 1 in the

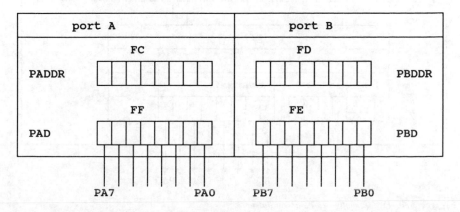

Figure 5.45 *The PIA direction and data registers*

DDR will cause the corresponding data line to be an output line. A 0 will cause it to be an input line.

For example, let us configure port A as an output port and port B as an input port.

LDA FF	Load ACC with FF, *i.e.*, 1111 1111
STA FC	Store ACC in A003, *i.e.*, PADDR
LDA 00	Load ACC with 00, *i.e.*, 0000 0000
STA FD	Store ACC in A002, *i.e.*, PBDDR

External data can then be placed on the lines PB0 to PB7 and read from address location FE. Similarly, any data placed in address FF will appear on the lines PA0 to PA7 (Figure 5.46). The control lines (CA1, CA2, CB1, CB2) provide the ability to co-ordinate the data transfer to and from external devices.

5.11.3 External Interfaces

So far we have been inside the computer. It is now time to look at the external environment and consider the issue of standards (Figure 5.47). For a standard to be acceptable it must provide the following features: completeness of electrical, physical, and logical definitions; flexibility to allow the connect of a wide range of devices; simplicity of use; transparency (*i.e.*, any data pattern can be transmitted); and a reliable method of data transfer. In addition, the standard must be internationally recognized and supported by the appropriate agencies.

5.11.3.1 External Interfaces—the Serial Link. A lot of interface and communications equipment send and receive data serially. Several standards exist for serial interfacing of which one will be considered in detail here, the RS series (Figure 5.48).

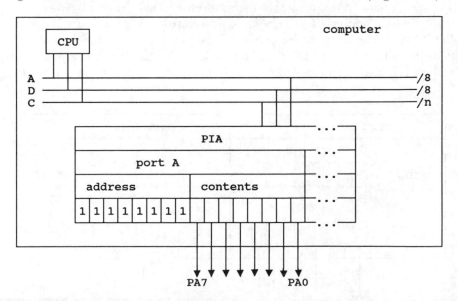

Figure 5.46 *Port A configured as an output port*

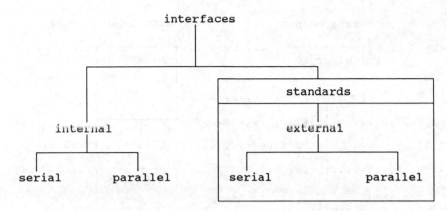

Figure 5.47 *External interfaces*

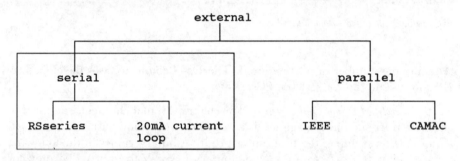

Figure 5.48 *External serial links*

The EIA has produced a variety of serial interface standards, the best known being the EIA-RS-232-C (EIA, Electronics Industry Association; RS, recommended standard; 232, standard number 232; C, revision C).

The RS-232-C standard, also referred to as simply RS 232, was designed for the single purpose that is clearly stated in the title of the specification: 'Interface between data terminal equipment (DTE) and data communications equipment (DCE) employing serial binary data interchange'. The limitations of the original standard resulted in various revisions. The current revision RS-232-D (1986) brings this interface into line with the international standards CCITT V.24 and others. It is used for both communications and instrumentation equipment. The standard defines the mechanical characteristics of the interface, the electrical signal characteristics, and the functional description of the interface circuits.

5.11.3.2 Mechanical Characteristics of the Interface. The mechanical characteristics of the interface refer to the interface between the DTE and the DCE. In particular, there must be a plug and a receptacle (socket) with the receptacle on the DCE. The pin number assignments are specified; however, the connector (DB-25, 25 pin, D type) that is specified in the RS-232-D was derived from the ISO standard. On some RS 232 compatible equipment, the DB-9 connector (9 pin, D type) is often used where there is no need for a 'handshake' protocol thereby

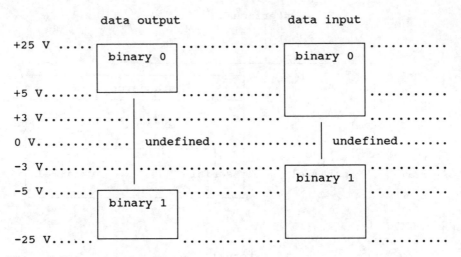

Figure 5.49 *Input and output data voltages*

reducing the number of pins required. The DB-9 connector was developed by IBM for use on the AT series of PCs.

5.11.3.3 Electrical Signal Characteristics. The electrical signal characteristics refer to characteristics such as voltage levels and grounding, along with the associated circuitry. For example, the transmitter produces binary values in the following ranges: binary 0, + 5 V to + 25 V; and binary 1, − 5 V to − 25 V. To allow for voltage drops over the line the data input values can be lower (Figure 5.49). Note that the voltage levels of microprocessors are typically from − 5 to + 5 V. It is therefore necessary to have a 'line driver' at the transmitting end to amplify the values to those required by the RS 232 specification. Many PCs have a standard + 15 to − 15 V output, which is within the RS 232 specification and can be used for the line driver. Similarly, at the receiving end, the higher voltages must be reduced to those appropriate to the microprocessor. The use of large voltage differences in the RS 232 standard ensures good noise immunity and allows data to be transmitted to up to about 15 m at 20 kHz.

5.11.3.4 Functional Description of the Interface Circuit. The functional description of the interface circuit defines the data, timing, and control signals used at the DTE to DCE interface (Figure 5.50). There are 25 electrical connections, many of which are not often used, that can be divided into the following four functional groups set into a 25-pin D connector (Figure 5.51): data lines; control lines; timing lines; and special secondary functions.

The data lines are used to transfer data using pins 2 and 3 (Figure 5.52). The data-flow reference point is the DTE interface, hence the following connections shown in Figure 5.53. At a functional level, there can be two separate wires representing two data paths, one for transmission and one for reception (Figure 5.53). The data signal is measured with respect to a common earth.

The control lines are used for interactive device control, which is often referred

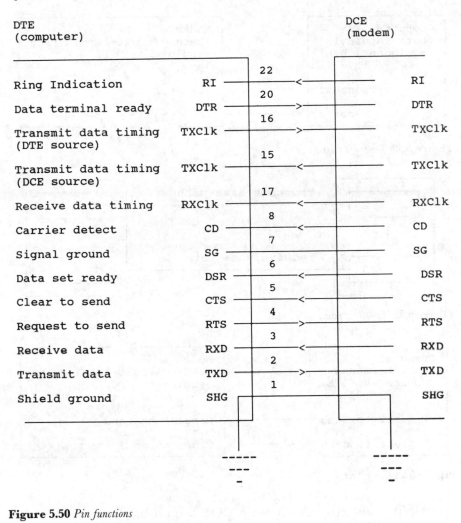

DTE
(computer)

DCE
(modem)

Ring Indication	RI	22 —<—	RI
Data terminal ready	DTR	20 —>—	DTR
Transmit data timing (DTE source)	TXClk	16 —>—	TXClk
Transmit data timing (DCE source)	TXClk	15 —<—	TXClk
Receive data timing	RXClk	17 —<—	RXClk
Carrier detect	CD	8 —<—	CD
Signal ground	SG	7	SG
Data set ready	DSR	6 —<—	DSR
Clear to send	CTS	5 —<—	CTS
Request to send	RTS	4 —>—	RTS
Receive data	RXD	3 —<—	RXD
Transmit data	TXD	2 —>—	TXD
Shield ground	SHG	1	SHG

Figure 5.50 *Pin functions*

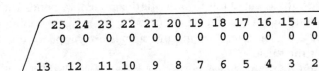

25	24	23	22	21	20	19	18	17	16	15	14	
0	0	0	0	0	0	0	0	0	0	0	0	
13	12	11	10	9	8	7	6	5	4	3	2	1
0	0	0	0	0	0	0	0	0	0	0	0	0

Figure 5.51 *The 25 pin D connector*

to as 'handshaking' (Figure 5.54). The most commonly used control lines used
for this purpose are request to send (RTS), clear to send (CTS), data set ready
(DSR, *i.e.*, DCE ready), and data terminal ready (DTR, *i.e.*, DTE ready). Note
that the handshake lines operate in reverse logic to the data lines, *i.e.*, a positive
voltage is binary 1 and a negative voltage is binary 0. These lines are used for half
duplex communication to control who is talking and who is listening. For

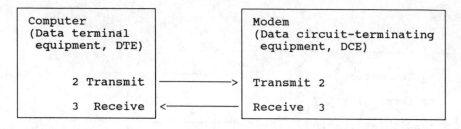

Figure 5.52 *Transmit/receive*

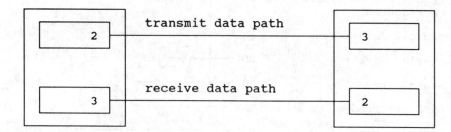

Figure 5.53 *Two data paths*

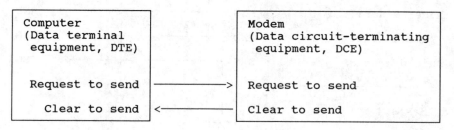

Figure 5.54 *Control lines*

example, the data terminal indicates RTS to the modem in order to transmit a character, but the data terminal must wait until the modem responds with CTS. The modem can only send CTS if it is not currently receiving a character. The ground connections are of note. Two ground signals are supplied, one for the signal and one for the shield.

The specification does not, however, define a protocol. The RS 232 is best suited to short (up to about 20 m), low-speed (up to about 20 kilobits s^{-1}) serial communication. The common data transmission rates are 110, 300, 600, 1200, 2400, 4800, 9600, and 19 200 baud. Although it can be used for both synchronous or asynchronous transmission, the RS 232 is typically used asynchronously.

5.11.3.5 Modems. Data communications equipment is a generic name for any equipment for the attachment of user devices to a network, for example, a modem (Figure 5.55). First we need to examine the modulator/demodulator (Figure 5.56). The normal telephone network was designed to transmit the human voice cheaply. To achieve some cost saving, the full range of the voice is

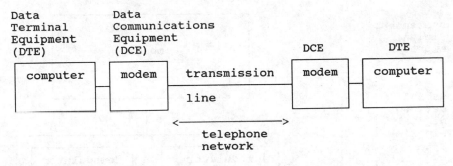

Figure 5.55 *DTE and DCE*

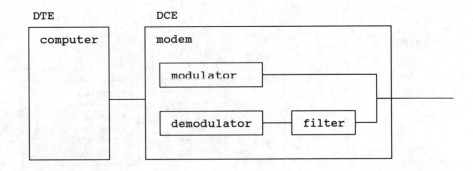

Figure 5.56 *The modem*

not transmitted, only up to 3 kHz. With this bandwidth it is not possible to send square waves as the distortion would be too great. Our digital signals are therefore converted by the modem into signals at frequencies below this 3 kHz limit. Called *frequency shift keying*, this involves converting each logic level into a frequency. Logic 1 and logic 0 are represented by 1270 and 1070 and also 2225 and 2025 Hz, respectively (Figure 5.57). At the receiving end, the frequencies are converted into logical signals. Data terminal equipment is a generic name for any user-supplied device connected to a network, for example, a computer.

The modulator converts the digital data into frequencies that are then transmitted down the telephone line. The receiving demodulator converts the frequencies into digital data. The filter on the demodulator removes frequencies generated by its own modem. This allows full duplex transmission. Two types of modem are in common use, the acoustic coupler (*i.e.*, a handset cradle), and a direct connection from the computer.

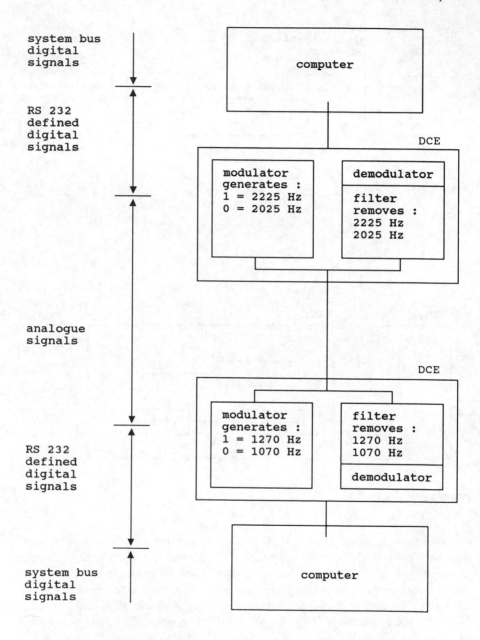

Figure 5.57 *Modem details*

The transmit data (TXD) transmits data and the receive data (RXD) receives data. The other lines are responsible for setting up the line, timing, and closing down or clearing the line. To illustrate this, let us consider how a call is established and a half duplex data exchange takes place. In this example, device A calls device B in order to send data (Figure 5.58).

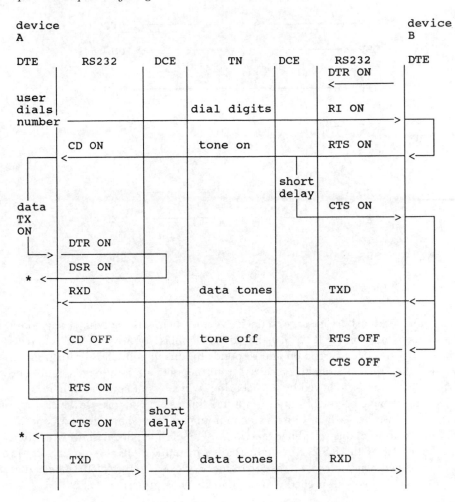

Followed by call clearing

Figure 5.58 *Connection protocol*

The connection must first be established by the user (device A) dialling the number of the remote computer (device B). Autodial is a possibility. If the remote device is free and ready to communicate, the ringing tone will stop and a single audio tone will be sent from device B to device A. The user of device A presses the data transmit button on the handset. This causes device A to be connected to the set up line. The success of this operation is indicated by the DSR lamp being turned on—a link has been established with the remote device B. When the number is dialled, the local modem at device B sets the ring indicator (RI) on and if the device is ready to accept the call (*i.e.*, the DTR line is on) the RTS line is turned on. This has the effect of acknowledging the incoming call by means of a single audio tone. After a short delay, which allows device A to prepare to receive data, the CTS is turned on and data can now be sent. Device B

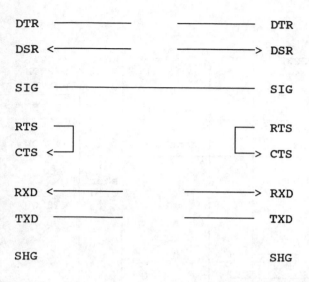

Figure 5.59 *The null modem*

may now send a short message to device A and then gets ready to receive data again by turning off the RTS (*i.e.*, the carrier signal is stopped). Device A detects the carrier has been turned off and responds by turning off CD, and turning on RTS. When the CTS signal is received from the modem, data may be sent. The whole procedure is followed by line clearing, which is not considered here.

So far we have been looking at computers linked by modems. However, there are many instances when we wish to communicate over short distances and do not need to use the telephone network. These short, local links are often implemented by the RS 232. A slight change in connections is required to form what is called a *null modem*, *i.e.*, one that does not exist (Figure 5.59). The actual number of connections needed varies, so do check your user manual.

5.11.3.6 Addressing the limitations of the RS 232. The main limitations of the RS 232 specification are as follows: the 'point-to-point' configuration is a restriction when there are a lot of 'intelligent' instruments to be connected together; the transmission distance is too low for many applications; the bit rate is too low for many applications; and the voltage specification is not directly compatible with microprocessors. The EIA has therefore developed a number of other interface standards to address these limitations. The RS 422 and RS 485 standard are particularly important for instrumentation. The RS 422 standard defines a differential or 'balanced' data communications interface and as such there must be two signal wires for each direction (Figure 5.60). Each data path must use two wires and the data signal is the difference in voltage across these two wires rather than in reference to a common earth wire. This two-wire differential voltage minimizes the effects of noise, as these effects are induced equally in both wires (Figure 5.61). Furthermore, the effects of voltage drops are minimized. The use of differential mode allows the use of lower voltages for the logic levels. To represent binary 0, the voltage on A is greater (by 5 V) than the voltage on B and, vice versa for binary 1 (Figure 5.62). Only one line driver is allowed per

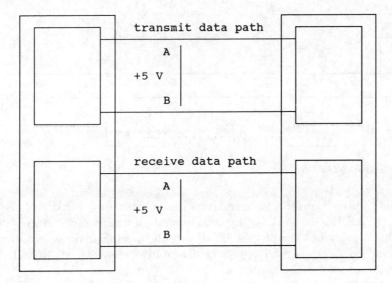

Figure 5.60 *The RS 422*

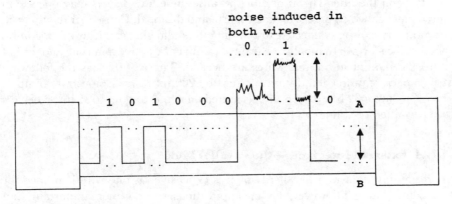

Figure 5.61 *Noise*

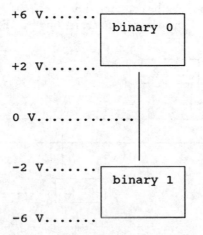

Figure 5.62 *RS 422 voltage levels*

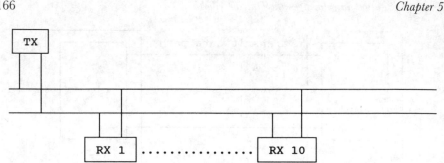

Figure 5.63 *RS 422 multiple receivers*

data line; however, up to 10 line receivers can be driven by that driver (Figure 5.63). The RS 422 has the following advantages: transmission distances of up to 1200 m; transmission rates of up to 10 Mbps; only one line driver is permitted on a line; and up to 10 line receivers can be driven by one line driver. Note that for full duplex, four wires are needed.

The RS 485 standard is based on the RS 422, but allows more transmitters and receivers on the line. This is possible because these line drivers may operate in three states called tri-states: logic 1, logic 0, and disabled. Consequently, up to 32 transmitters can be connected to the same line allowing 'multidrop' operation (Figure 5.64). Note that only one transmitter may be active at a time and for full duplex communication, four wires are needed. The RS 485 has the following advantages: transmission distances of up to 1200 m; transmission rates of up to 10 Mbps; up to 32 line drivers are permitted on a line; and up to 32 line receivers are permitted on a line.

5.11.4 External Interfaces—the Parallel Link

Transmitting data in a parallel fashion is expensive, as one wire is needed for each data channel; however, in some applications, speed is more important than cost (Figure 5.65). There are a variety of parallel interfacing standards one of which, the IEEE 488, will be considered here.

Standardization of the Hewlett Packard bus by IEEE 488–1975 (standard

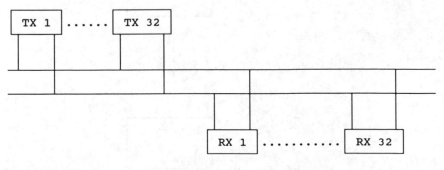

Figure 5.64 *RS 485*

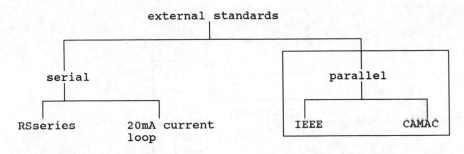

Figure 5.65 *External standards*

digital interface for programmable instrumentation) and the ANSI standard MC1.1 1975 has encouraged the use of this method for parallel interfacing. The method was designed 'to provide an effective communication link over which messages are carried in an unambiguous way among a group of interconnected devices'.

The IEEE interface design allows the simultaneous connection of up to 15 devices or instruments on a common parallel communications bus. In order to differentiate between them, each instrument is allocated a specific address between decimal 0 and 30 by selecting that address on a set of switches at the rear of each instrument. The standard defines methods for the orderly transfer of data, the correct addressing of individual units, bus management commands, and the physical details of the connectors. The bus may be up to 20 m long and operates asynchronously with a data rate of up to 1 Mbyte s^{-1}. In this asynchronous mode, the limiting factor to the data transfer rate is usually the devices being used. The bus uses a three-wire handshake to transmit bit parallel, byte serial data. The bus devices can be arranged into the following groups: measurement devices (*e.g.*, spectrum analysers and digital multimeters); stimuli devices (*e.g.*, pulse generators); those that provide analogue or digital stimuli or power to the unit under test (*e.g.*, d.c. power supplies); output devices (*e.g.*, printers and *x–y* plotters); storage devices (*e.g.*, floppy-disk drives and hard-disk drives); input devices (*e.g.*, digitizers); and other devices (*e.g.*, filters and IEEE bus extenders).

5.11.4.1 Bus Configurations. The devices on the IEEE 488 may be connected in a star or linear-chain configuration. In the star configuration, each instrument is connected to the controller by means of its own cable (Figure 5.66). One restriction of this configuration is that due to the limitation of the maximum cable length, all instruments must be close to the controller. In the chain configuration, each device is connected to the adjacent one. The controller does not have to be the first or last in the chain, it can be placed anywhere (Figure 5.67). This is the most typical configuration. These are the suggested configurations; however, any connections are possible provided the following criteria are met: there must be no more than 15 devices (including the controller) on the bus; the cable length between adjacent devices must not exceed 4 m; the average

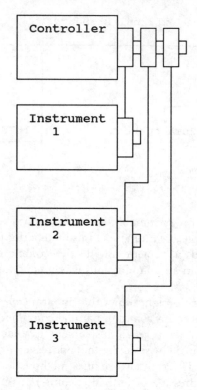

Figure 5.66 *IEEE 488 star configuration*

separation of devices over the entire bus must be 2 m; and the total cable length must not exceed 20 m.

5.11.4.2 Device Types. There are four classes of device specified: listeners, talkers, talkers/listeners, and controllers.

A listener is a one-way communication device that can only receive data from other devices. It cannot send data. It receives instructions from the controller, which determines when the data are to be read. For example, printers are listeners.

A talker is a one-way communication instrument or device that can only send data to other devices. It cannot receive data but it can receive instructions from the controller, which determines when the data is to be transmitted. For example, analogue-to-digital converters are talkers.

A talker/listener has the characteristics of both talkers and listeners. However, both functions cannot operate at the same time. This is the most flexible configuration that is suitable for intelligent instrumentation. Data parameters can be sent to the instrument, and then the instrument can act as a data logger.

A controller manages the entire bus. It is typically a PC or a microprocessor-based instrument. The controller determines which devices will send data, which devices will receive data and when. There can only be one active controller, the

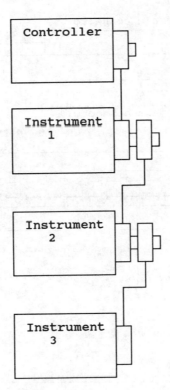

Figure 5.67 *IEEE 488 chain configuration*

controller in charge (CIC). There may be several controllers; however, only one can be active at a time.

5.11.4.3 Mechanical and Electrical Specification. The IEEE 488 uses a shielded 24-wire cable and standard Amphenol connectors allowing stacking and connection security with screws (Figure 5.68).

5.11.4.4 Bus Structure of the IEEE 488. There are eight bidirectional data lines and eight data management lines (three data transfer control lines and five general data management lines) (Figure 5.69). Each device is connected to the bus structure (Figure 5.70). At any point in time, the controller will determine which individual device can be the talker and which device or devices must listen. The controller therefore determines the origin of the data and its destination(s). There can only be one talker but there may be one or more listeners at any instance in time. The data bus has a dual role; it is used by the controller to send instructions to other talkers and listeners, it is then freed by the controller so that data may be sent on it.

5.11.4.5 Bus Handshake Control Lines. To illustrate how the IEEE 488 bus works, let us first consider the bus handshake control lines. Three lines are used to set up a handshake sequence for asynchronous operation. The key to understanding the IEEE bus is understanding the inverse logic of the three bus handshake control

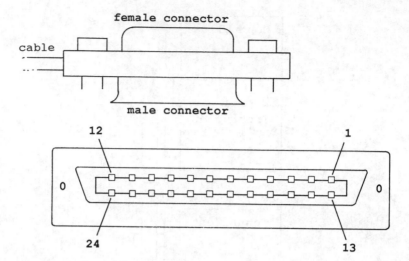

Figure 5.68 *IEEE 488 mechanical specification*

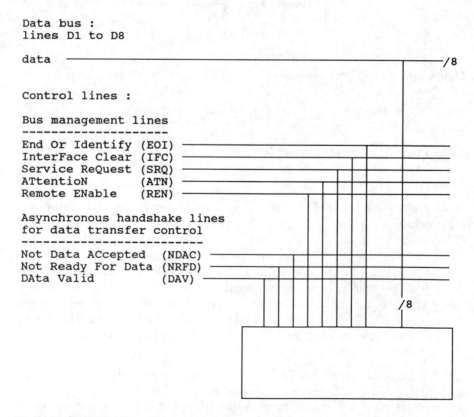

Figure 5.69 *IEEE 488 bus structure*

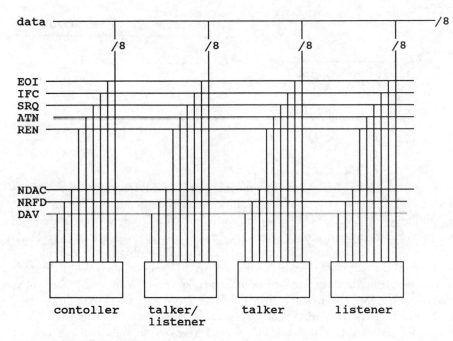

Figure 5.70 *IEEE 488 device connection*

lines, not ready for data (NRFD), data available (DAV), and not data accepted (NDAC).

When the NRFD line is low this means that it is TRUE that 'all the acceptors ARE NOT ready for data'. When the line goes high it means that the statement 'all the acceptors ARE NOT ready for data' is FALSE—which means that 'all the acceptors ARE ready for data' (Figure 5.71).

When the DAV line is low this means that it is TRUE that 'data ARE available'. When the line goes high it means that the statement 'data ARE available' is FALSE—which means that 'data ARE NOT available' (Figure 5.72).

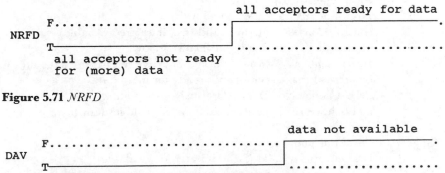

Figure 5.71 *NRFD*

Figure 5.72 *DAV*

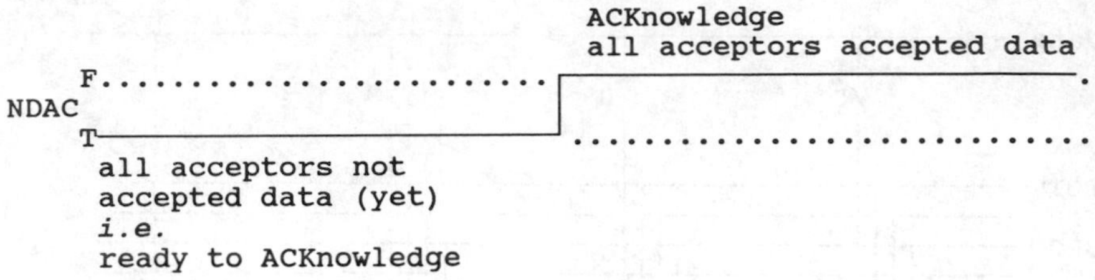

```
                                    ACKnowledge
                                    all acceptors accepted data
          F........................┌─────────────────────────────────.
NDAC                               │
          T────────────────────────┘ ...........................
          all acceptors not
          accepted data (yet)
          i.e.
          ready to ACKnowledge
```

Figure 5.73 *NDAC*

When the NDAC line is low this means that it is TRUE that 'all acceptors HAVE NOT accepted data'. When the line goes high it means that the statement 'all acceptors (HAVE) NOT accepted data' is FALSE, which means that 'all acceptors HAVE accepted data'. Another way of looking at this is that the NDAC line acts like an ACKnowledgement (ACK) line. When high, this provides an ACK signal that the data have been accepted by all the acceptors. When the data are removed from the data line the line must be taken down ready to ACK the next data (Figure 5.73). The master control line is the DAV line. The talker places data on the bus and asserts the DAV line to signal its validity to the listeners.

The two slave control lines are NDAC and NRFD. The NRFD line is held at logic 0 until all devices commanded to listen are ready. The NDAC line is held at logic 0 by all devices commanded to listen until they have successfully received the data. The line only becomes logic 1 when the last (and slowest reading) device has received the data. The basic data transfer mechanism is simply and automatically adjusts to the data rate of the slowest device. The sequence is shown in Figure 5.74. The controller is talking and is referred to as the SOURCE. It controls the DAV line while the listener, called the ACCEPTOR, controls the NRFD and NDAC lines. The names and lines each device controls are reversed when their respective listen and talk roles are reversed. The handshake sequence is as follows:

Step 1	NRFD goes high
	All listeners ready for data
Step 2	DAV line (DAta Valid) goes low
	controller asserts data are valid (*i.e.*, data are available)
Step 3	NRFD lines (Not Ready For Data) go low
	Instruments say, do not change data while current data are being read (*i.e.*, acceptors not ready for more data)
Step 4	NDAC line (Not Data ACcepted) goes high when all instruments have read the data (*i.e.*, ACKnowldge data have been read)
Step 5	DAV line goes high
	Data no longer valid (*i.e.*, data are not available)
Step 6	NDAC line goes low
	Instruments remove data accepted indication
Step 7	Step 1 again

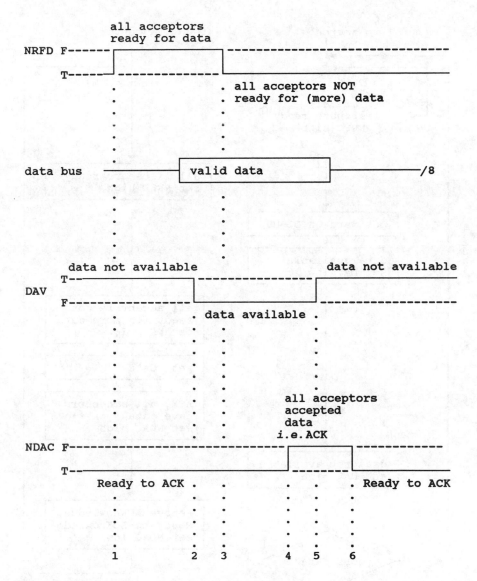

Figure 5.74 *Handshaking*

It can sometimes help to see this as a co-ordinated flow chart (Figure 5.75).

The IEEE 488 offers a range of complex functions, but here we will only consider the basic principles. One controller must be specified as the system controller having control of the interface clear (IFC) and remote enable (REN) signal lines. These permit an orderly start up of the bus and also recovery in case of failure. The REN line going low causes all the instruments on the bus to go into the remote state. The IFC then pulses low to stop any bus activity. The ATN line can then be used by the current controller device to inform devices on the

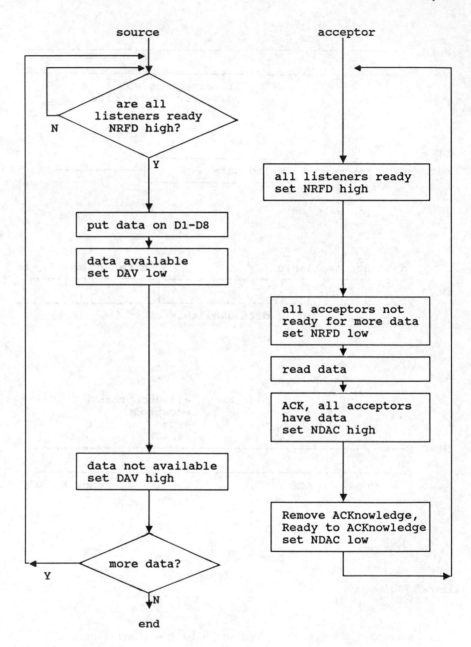

Figure 5.75 *Handshake flowchart*

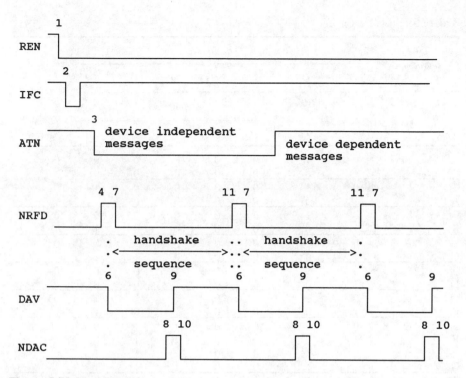

Figure 5.76 *Line initialization*

bus that the data transfer is control information that may affect them (Figure 5.76). If the ATN line is low, the DB holds messages used to control the interface system itself, *i.e.*, interface messages. If the ATN is high, the DB holds messages used by the devices connected to the bus, *i.e.*, device-dependent messages. Data (device-dependent or independent) are then transferred using the handshake lines.

If a device needs service, then it sets the service request (SRQ) true. This allows the device to request attention only when it is ready. The controller can poll all the devices in series to find out which devices need servicing. To carry out this serial poll, the controller unlistens all devices and then issues the serial poll enable (SPE) command. This addresses each device sequentially (ATN true and talker address) and in turn receives the status (ATN false, device sends status). When the poll is complete the serial poll disable (SPD) terminates the polling. It is possible to conduct a parallel poll.

5.11.4.6 Addresses. Each instrument may be set up for any address between 0 and 30 (decimal). Referring to Table 5.1 it can be seen that there are two ASCII characters corresponding to each decimal address depending on whether bit D6 or bit D7 is a 1. These two characters determine whether the instrument is to talk, or listen when addressed. For example, assume that the controller has address 21 and that it is going to set up an intelligent pH meter to measure the

Table 5.1 *Address table*

Talk	Listen	Talk	Listen	16	8	4	2	1	Decimal address
	SP	0	1	0	0	0	0	0	00
@		1	0	0	0	0	0	0	
	!	0	1	0	0	0	0	1	01
A		1	0	0	0	0	0	1	
	"	0	1	0	0	0	1	0	02
B		1	0	0	0	0	0	1	
	#	0	1	0	0	0	1	0	03
C		1	0	0	0	0	1	1	
	$	0	1	0	0	1	0	0	04
D		1	0	0	0	1	0	0	
	%	0	1	0	0	1	0	1	05
E		1	0	0	0	1	0	1	
	&	0	1	0	0	1	1	0	06
F		1	0	0	0	1	1	0	
		0	1	0	0	1	1	1	07
G		1	1	0	0	1	1	1	
	(	0	1	0	1	0	0	0	08
H		1	0	0	1	0	0	0	
	)	0	1	0	1	0	0	1	09
I		1	0	0	1	0	0	1	
	*	0	1	0	1	0	1	0	10
J		1	0	0	1	0	1	0	
	+	0	1	0	1	0	1	0	11
K		1	0	0	1	0	1	0	
	'	0	1	0	1	0	1	1	12
L		1	0	0	1	0	1	1	
	−	0	1	0	1	1	0	0	13
M		1	0	0	1	1	0	0	
	.	0	1	0	1	1	0	1	14
N		1	0	0	1	1	0	1	
	/	0	1	0	1	1	1	0	15
O		1	0	0	1	1	1	0	
	0	0	1	0	1	1	1	1	16
P		1	0	0	1	1	1	1	
	1	0	1	1	0	0	0	0	17
Q		1	0	1	0	0	0	0	
	2	0	1	1	0	0	0	1	18
R		1	0	1	0	0	0	1	
	3	0	1	1	0	0	1	0	19
S		1	0	1	0	0	1	0	
	4	0	1	1	0	0	1	1	20
T		1	0	1	0	0	1	1	
	5	0	1	1	0	1	0	0	21
U		1	0	1	0	1	0	0	

TALK	LISTEN	D7	D6	D5	D4	D3	D2	D1	DECIMAL
6		0	1	1	0	1	0	1	22
	V	1	0	1	0	1	0	1	
7		0	1	1	0	1	1	0	23
	W	1	0	1	0	1	1	0	
8		0	1	1	0	1	1	1	24
	X	1	0	1	0	1	1	1	
9		0	1	1	1	0	0	1	25
	Y	1	0	1	1	0	0	1	
:		0	1	1	1	0	1	0	26
	Z	1	0	1	1	0	1	0	
;		0	1	1	1	0	1	1	27
	[	1	0	1	1	0	1	1	
<		0	1	1	1	1	0	0	28
	\	1	0	1	1	1	0	0	
=		0	1	1	1	1	0	1	29
	]	1	0	1	1	1	0	1	
>		0	1	1	1	1	1	0	30
	^	1	0	1	1	1	1	0	

pH and return the value. A pH meter set to address 04 would know that it was being addressed to talk when it detected a 'D' on the bus, and would respond as a listener when it detected a '$' (Figure 5.77).

5.11.4.7 Bringing Everything Together. Assume that the controller has address 21 and that it is going to set up an intelligent pH meter to measure the pH and return the value. The pH meter is set to address 04. When the controller is talking it is referred to as the SOURCE and it controls the DAV line; the pH

ASCII characters		Data lines							
TALK	LISTEN	D7	D6	D5	D4	D3	D2	D1	DECIMAL ADDRESS
TALK	LISTEN	TALK	LISTEN	ADDRESSES					
				16	8	4	2	1	
.	.	.	.	.	.	.	.	.	.
D	$	0 / 1	1 / 0	0 / 0	0 / 0	1 / 1	0 / 0	0 / 0	04
.	.	.	.	.	.	.	.	.	.
U	5	0 / 1	1 / 0	1 / 1	0 / 0	1 / 1	0 / 0	0 / 0	21
.	.	.	.	.	.	.	.	.	.

Figure 5.77 *Devices addresses*

18 M, programs off
19 0
20 T, internal trigger
21 0
22 ? terminator to force pH meter to execute the above commands and take a pH reading (*i.e.*, pH on)
23 CR (Carriage Return) indicates the end of the data message to the instrument
24 LF (Line Feed)
25 ATN low. Following is an interface message
26 ? on data bus, *i.e.*, unlisten again, all instrument to respond
27 D on data bus
     Controller informs pH meter that it must talk. Only the pH meter recognizes this and can respond
28 5 on bus, controller's own listener address
29 ATN high, end of interface message
     The pH meter can now send its own message
30 7
31 .
32 5

This sequence can be shown on a timing chart (Figure 5.78).

Whilst this may seem a long sequence of events, it is a reliable method for communication between intelligent instruments. The SRQ line is used by instruments to indicate to the controller that they require service. The instruments may be programmed to request service when certain conditions arise, *e.g.*, fault occurrence. The controller can then do either a serial or parallel poll to determine which instrument sent the SRQ, and act accordingly. This and other commands will not be further considered here; however, a complete address table is included (Table 5.1).

5.12 DIRECT MEMORY ADDRESSING

In programmed data transfer (PDT) all data transfers between the CPU and the I/O device are completely under program control. The CPU typically has four types of I/O instructions used to transfer data, read data from device, write data to device, set individual control flags in interface unit, and test individual control flags in interface unit. The CPU addresses the interface unit (IU) and tests the status register (SR). If the device is ready, there is data transfer.

To illustrate, let us consider a small piece of assembly code to perform a data transfer operation (Figure 5.79). With a 1 MHz clock speed and assuming two clock cycles per instruction the time taken for this fragment of code to transfer 1 byte of data is 10 μs. This gives a data transfer rate of $1/10 = 100\,000$ bytes s^{-1}. As we have seen, a high-performance disk has data transfer rates of 625 000 bytes s^{-1}. It may be necessary to transfer large quantities of data between very fast I/O devices. In such a case, the processor could not keep up! The limiting factor in this type of data transfer is the time taken for the processor to execute the small

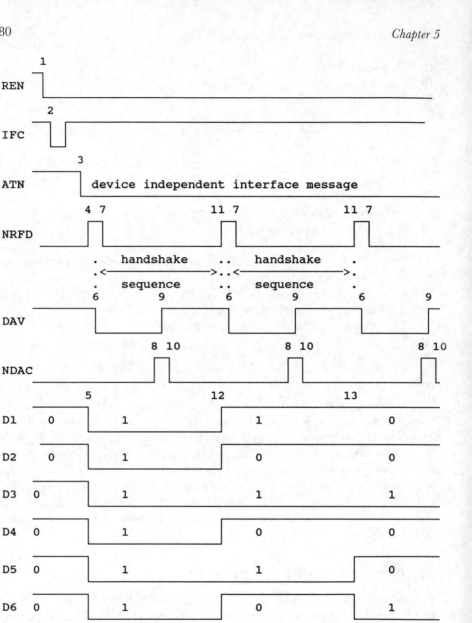

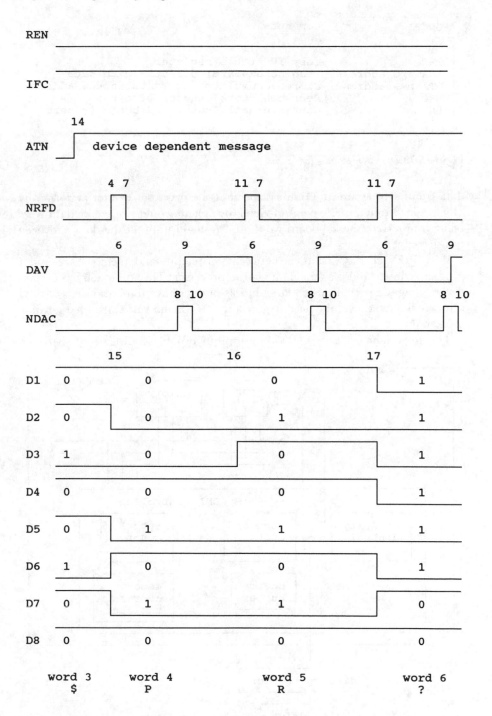

Figure 5.78 *IEEE 488 Timing chart*

```
    Code                    Comments

 ┌──▼
 │  TST                     Test IU status register
 │  LDA I/O address         Load accumulator directly with data
 │  STA new address         Store accumulator contents in new address
 │  DEC                     Decrement total number of bytes
 │  BNE                     Continue until number of bytes is zero
 └──
```

Figure 5.79 *Data transfer program*

data transfer program, the result being that the processor cannot perform any other tasks. From a user's point of view, the system would stop while data were being transferred; the keyboard would not respond until all the data had been transferred.

With PDT all data transfers between peripherals and memory are under program control and pass the data via the processor. This approach has certain disadvantages especially when large blocks of data have to be transferred: data rates are limited due to the need to go via the ACC; and valuable processor time is taken up.

The direct memory access (DMA) technique provides a direct route between

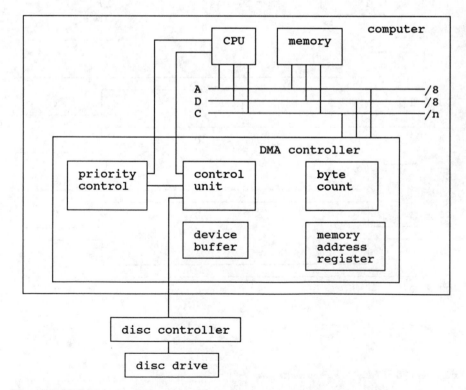

Figure 5.80 *DMA controller*

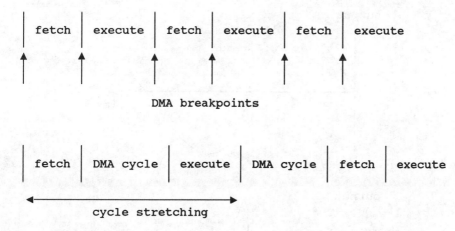

Figure 5.81 *Cycle stealing*

I/O devices, such as disk drives, and the primary memory of the computer. There is no need for continuous CPU involvement, as there is a special controller dedicated to DMA (Figure 5.80). In this arrangement, hardware facilities are included in the CPU that enable the DMA device to insert or extract data directly to or from the computer memory, bypassing all program control after initialization. This technique is suited to the transfer of large data blocks in a single continuous operation.

To initialize a DMA transfer, typically the following information is transferred to the DMA control unit: the peripheral selection; the direction of transfer; the memory start address; and the number of bytes (or words) to be transferred.

After this initialization, the processor is free to execute any program, and each time the selected peripheral device is ready to transfer a word to or from the memory a 'hesitation' takes place. The DMA process 'steals' a memory cycle from the program currently being executed. The word is then transferred (Figure 5.81). Note that a DMA grant will only be given at certain stages within the fetch/execute cycle.

5.13 ANALOGUE TO DIGITAL AND BACK AGAIN

In the 'real world' there are *analogue* signals. There are various types of analogue signal, but typically they are continuous and can take any value within pre-defined limits (Figure 5.82). By contrast, digital signals have only predefined acceptable values (Figure 5.83). Values outside these predefined limits are not used. For the computer to interact and control an analogue environment, there must be a conversion between digital and analogue signals (Figure 5.84). For the biologically minded, it is interesting to note that our senses also perform this conversion. Sensory organs stimulate neurons to transmit electrical pulses. Each pulse has digital characteristics. The two main methods of interfacing to a PC are to use either an external or internal bus product (Figure 5.85). An external bus

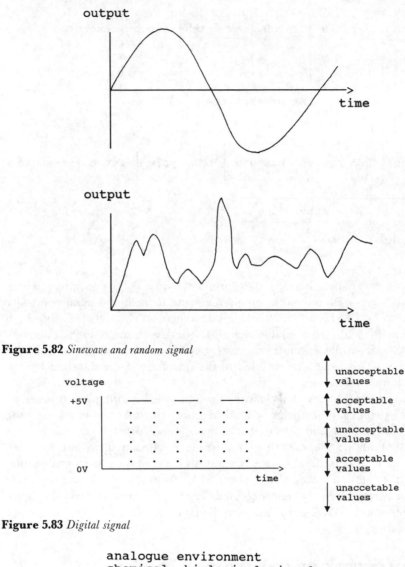

Figure 5.82 *Sinewave and random signal*

Figure 5.83 *Digital signal*

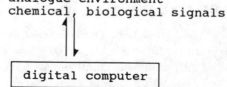

Figure 5.84 *Analogue environment*

product, such as the RS series, allows communication with a remote device, a data logger. The data logger is responsible for data collection and conversion from analogue to digital. The advantages of an external bus product include the following: remote data acquisition is possible; different computers may use the

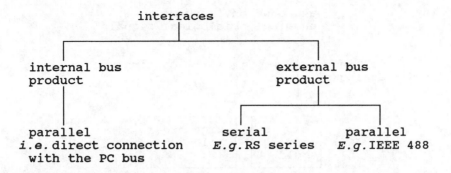

Figure 5.85 *Interfaces*

same data logger; there are no restrictions due to the PC bus size; and the data logger may be 'intelligent' and can preprocess the data, thereby relieving the PC of computational demands.

By contrast, an internal bus product is typically a p.c.b. that is mounted on the PC mother board connecting directly with the PC system bus. The advantages of an internal bus include the following: high speed; and possibly, lower costs.

However, regardless of the type of interface chosen, there are considerations that affect all types of interface. The analogue signal may be continuously changing. A *sample and hold* circuit is therefore needed to sample the input signal and then to hold the value (discrete analogue) while the analogue to digital converter (A/D converter or ADC) forms the associated digital signal. A digital to analogue converter (D/A converter or DAC) forms discrete analogue values from digital signals. The high-order hold circuit then forms the associated continuous analogue signal (Figure 5.86).

5.13.1 Sample and Hold Circuits

Because the input to an ADC must be quantized, it is often necessary, owing to the changing nature of the input, to sample the input periodically and to hold that value until the conversion is complete. This sampled analogue signal has gaps between the adjacent data sample points (Figure 5.87). With correct sampling, interpolation between these points allows an accurate representation of the original signal. However, problems can arise if the sampling frequency is inadequate. If the sampling frequency is too low, then some of the information contained in the input signal may be lost resulting in aliasing (Figure 5.88). The Nyquist criterion states that: 'The original signal can be completely recovered without distortion if it is sampled at a rate at least twice that of the highest frequency contained in the continuous bandwidth of the original signal'. In effect, the sampling frequency $\geq 2Fh$, where Fh is the highest frequency component of the signal being sampled.

An easy method for determining whether the sampling frequency is high enough is to perform a frequency analysis of the data signal. An analogue voltage signal can be sampled periodically using a fixed-frequency clock pulse train. At

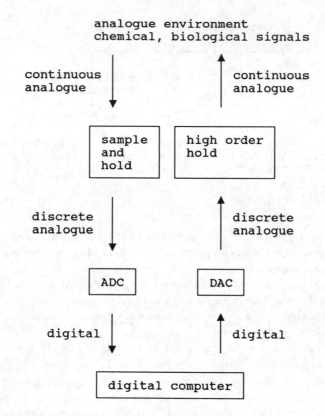

Figure 5.86 *Analogue to digital conversion*

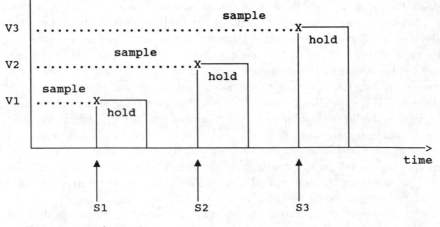

Figure 5.87 *Sample and hold values*

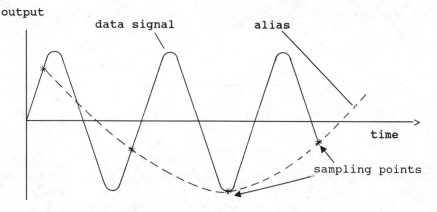

output

Figure 5.88 *Aliasing*

each clock pulse, the magnitude of the analogue signal at that particular instant in time is allowed to pass to the next stage of conditioning. We now have a sampled signal. This type of waveform is called return to zero (RTZ) (Figure 5.89). If, however, the value is stored after each pulse this can be converted to a non return to zero (NRZ) sampled waveform (Figure 5.90). The sample and hold circuit samples the analogue input signal at discrete intervals and the values at these intervals are passed to the ADC. When the conversion of one value has been completed, the same and hold (S/H) circuit takes another value. The complete S/H circuit has an input buffer to capture the analogue input signal. The switch actuator determines when the input signal will be transferred to the output circuit (Figure 5.91). Amongst others, the following parameters can be used to select the appropriate S/H chip: the aperture time, *i.e.*, the time between the hold command and the time at which the sample is taken; the aperture uncertainty time, *i.e.*, the difference between the maximum and minimum aperture times; and the acquisition time, *i.e.*, the time required for the output of the chip to equal the input value.

5.13.2 Analogue to Digital Converter

An ADC produces a digitally coded output. The digital value is proportional to the analogue voltage applied at the input. Our digital signal is not continuous, but has discrete levels. The size of this granularity depends upon the incremental step size, LSB. Figure 5.92 shows a 3 bit convertor, more typically we would use an 8 or 12 bit device. We could represent voltage as follows:

Voltage/V	Binary value
0	0000 0000
1	0000 0001
2	0000 0010
.	.
.	.
.	.
225	1111 1111

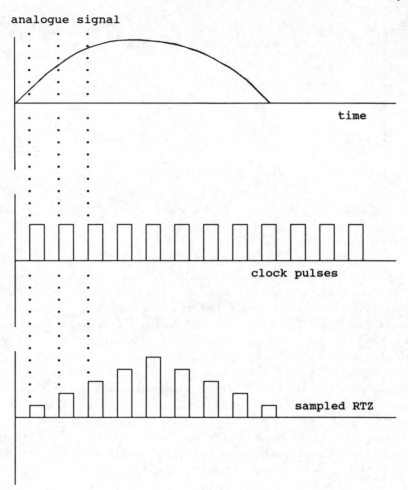

Figure 5.89 *Sampled analogue signal*

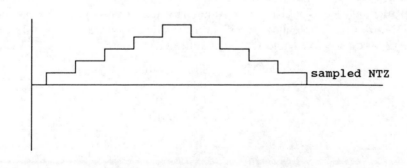

Figure 5.90 *Sample and hold signal (NRZ)*

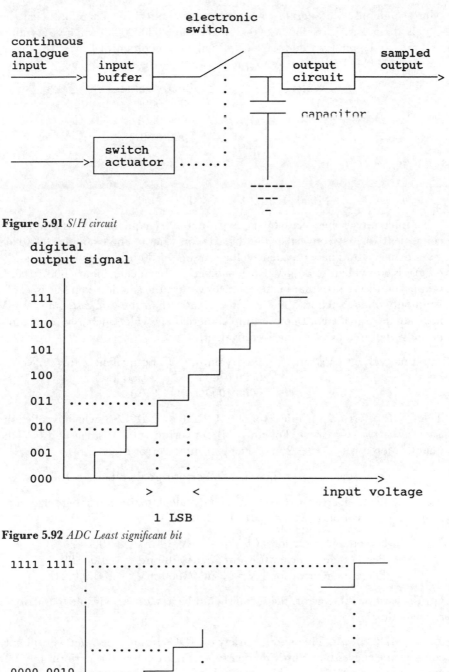

Figure 5.91 *S/H circuit*

Figure 5.92 *ADC Least significant bit*

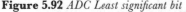

Figure 5.93 *Eight bit ADC*

More commonly we operate with a range of about 1 V. Therefore each step equals about 4 mV. In this case, the binary value 000000000 will give a voltage of 0 mV, and a binary value of 00000001 will give a voltage of 4 mV. The binary value of 11111111 will therefore represent 1.02 V (Figure 5.93).

Voltage/mV	Binary value
0	0000 0000
4	0000 0001
8	0000 0010
.	.
.	.
.	.
1020 (= 1.02 V)	1111 1111

To interconvert between voltage and the binary equivalent it is simple and convenient to first convert to the denary equivalent. For example, 00001111 gives denary 15. The equivalent voltage is 15×4.00 mV = 60 mV. Similarly, 80 mV is equivalent to 80 mV/4 mV = denary 20, which is binary 00010100. It is important to realize that in our 8 bit convertor, the smallest step size is 4 mV. An input voltage such as 6 mV is 2 mV greater than the 4 mV step and 2 mV less than the 8 mV step. In our example, therefore, a value such as 6 mV cannot be converted to an exact binary equivalent.

Input voltage (V_{in})/mV	Binary output	Equivalent voltage/mV
6	0000 0100	4 (too low)
	0000 1000	8 (too high)

There is therefore an intrinsic error of $-1/2$LSB to $+1/2$LSB, which is called the *quantization conversion error*. For an 8 bit convertor with a range of 1.0 V the quantization error is $+/-2$ mV. The percentage error is given by

$$\text{percentage error} = (\text{quantization error/signal voltage}) \times 100$$

For a quantization error of $+/-2$ mV, let us calculate the percentage error for a range of input voltages (V_{in}) = 0.01, 0.1, and 1.0 V.

Percentage error for 0.01 V = (2 mV/10 mV) × 100 = 20%
Percentage error for 0.1 V = (2 mV/100 mV) × 100 = 2%
Percentage error for 1 V = (2 mV/1000 mV) × 100 = 0.2%

For the least possible error, the signal should be as close as possible to the upper range of the device.

5.13.2.1 ADC circuits. There are a variety of ADC circuit types, but we will only consider the successive approximation type (Figure 5.94). To perform the A/D conversion, the ADC senses the magnitude of the input and compares it to some reference level (V_{ref}). If there is a difference in levels it measures the difference digitally (by a counting process). At the start conversion signal, the MSB of the DAC is activated. This produces 1/2 Full Scale Voltage (FSV) at the output of the DAC. If the output state of the comparator indicates that the input voltage (V_{in}) is greater than the DAC output, the DAC MSB is left on and the

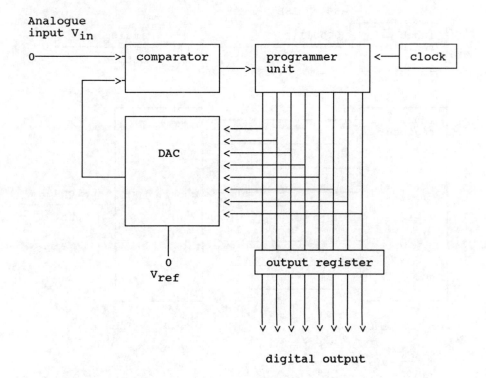

Figure 5.94 *ADC circuit*

programmer unit enables the next significant bit of the DAC. If, however, the output state of the comparator indicates that the DAC output is greater than V_{in}, the programmer unit turns off the DAC's MSB and enables the next most significant bit. This process continues with the programmer unit making successive decisions until the input to the output storage register (*i.e.*, the input to the DAC) is the digital equivalent of V_{in}. All this comparison takes time, so an ADC cannot produce an output as fast as a DAC. The speed of the DAC is limited by the response time of the circuit. In ADCs, the limiting factor is the counting time.

	Voltage contribution							
	MSB							LSB
Bit weight	1/2	1/4	1/8	1/16	1/32	1/64	1/128	1/256
1st decision	1							
2nd decision	1	1						
3rd decision	1	1	0					
4th decision	1	1	0	1				
5th decision	1	1	0	1	1			
6th decision	1	1	0	1	1	1		
7th decision	1	1	0	1	1	1	1	
8th decision	1	1	0	1	1	1	1	0
Binary	1	1	0	1	1	1	1	0

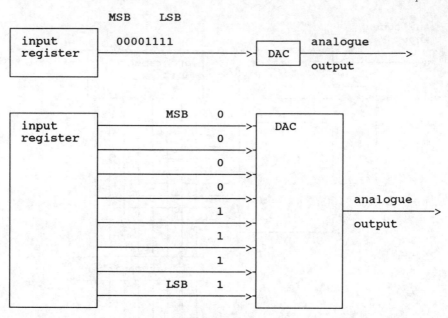

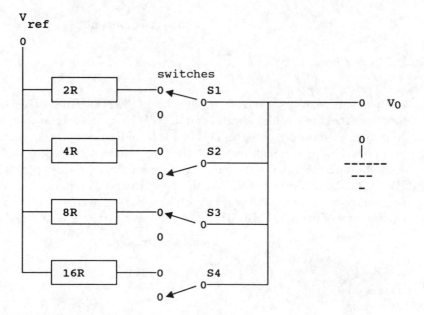

Figure 5.95 *Serial/parallel*

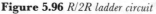

Figure 5.96 *R/2R ladder circuit*

5.13.3 Digital to Analogue Converter

The DAC produces an output voltage proportional to the input digital code. The digital input can be in either serial or parallel fashion (Figure 5.95). In the serial method, only one data input line is needed, but, for an 8 bit system, eight clock pulses are required to transfer the data. Only one clock pulse is needed for the parallel input; however, 8 data input lines are needed. Due to its speed of operation, the parallel DAC is the most common. There are different circuits that can be used. These circuits are available on a single chip. A common circuit is the R/2R ladder circuit (Figure 5.96).

S1	S2	S3	S4					
1/2	1/4	1/8	1/16					
0	0	0	0	0 + 0	+ 0	+ 0	−	0
0	0	0	1	0 + 0	+ 0	+ 1/16 =	1/16
0	0	1	0	0 + 0	+ 1/8 + 0	=	2/16
0	0	1	1	1 + 0	+ 1/8 + 1/16 =	3/16
0	1	0	0	0 + 1/4 + 0	+ 0	=	4/16
0	1	0	1	0 + 1/4 + 0	+ 1/16 =	5/16
0	1	1	0	0 + 1/4 + 1/8 + 0	=	6/16
0	1	1	1	0 + 1/4 + 1/8 + 1/16 =	7/16
1	0	0	0	1/2 + 0	+ 0	+ 0	=	8/16
1	0	0	1	1/2 + 0	+ 0	+ 1/16 =	9/16
1	0	1	0	1/2 + 0	+ 1/8 + 0	= 10/16
1	0	1	1	1/2 + 0	+ 1/8 + 1/16 = 11/16
1	1	0	0	1/2 + 1/4 + 0	+ 0	= 12/16
1	1	0	1	1/2 + 1/4 + 0	+ 1/16 = 13/16
1	1	1	0	1/2 + 1/4 + 1/8 + 0	= 14/16
1	1	1	1	1/2 + 1/4 + 1/8 + 1/16 = 15/16

$$V_0 = V_{ref}(1/2 + 1/4 + 1/8 + 1/16)$$

5.13.4 ADC and DAC Characteristics

Converters have the following different performance characteristics that must be considered, as shown in Figure 5.98: resolution and word length; maximum output, *i.e.*, full scale range (FSR); accuracy (quantization conversion error); conversion rate; stability; monotonicity; linearity; differential linearity; offset error; and gain or scale error.

5.13.4.1 Resolution and Word Length. The smallest quantum is 1 LSB. In voltage terms we have

$$\text{Resolution} = 1 \text{ LSB} = \text{voltage range}/2^n$$

where n is the converter word length. The resolution of an ADC, for example, is the smallest change in the analogue input that will cause a 1 bit change in the digital output. The resolution determines the minimum value difference that can produce an output.

5.13.4.2 Maximum Output—Full-scale Range. The maximum possible output voltage of the converter represents the digital value of all 1s.

$$\text{Full-scale range} = \text{voltage range} \times (2^n - 1)/2^n$$

where n is the word length. For example, the range of the full-scale range is the maximum value of analogue voltage that the DAC can output.

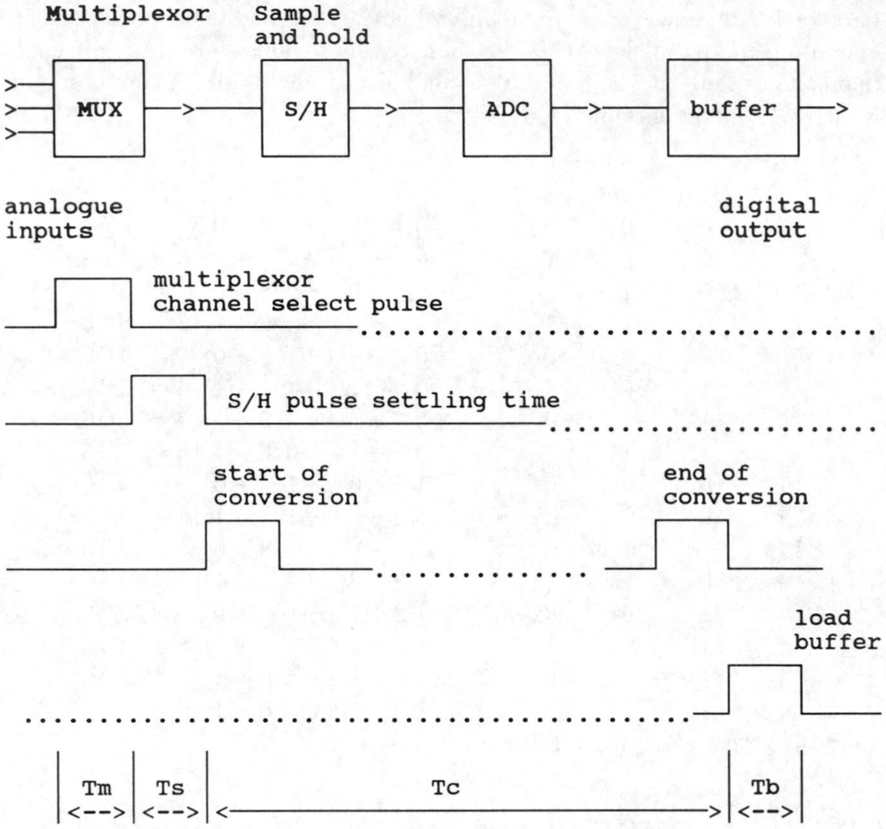

Conversion rate = 1/(Tm + Ts + Tc + Tb)

where

Tm : multiplexing switch time
Ts : S/H settling time
Tc : ADC conversion time
Tb : buffer time

Figure 5.97 *ADC conditioning circuit*

5.13.4.3 Accuracy (Percentage Quantization Error). The intrinsic error of − 1/2 LSB to + 1/2 LSB, which is called the *quantization conversion error*, is given by

percentage quantization error = (quantization error/signal voltage) × 100

5.13.4.4 Conversion Rate. This is not just a reciprocal of the conversion time. There are several considerations due to the fact that there are a number of elements and stages in the signal conversion. Consider, for example, an ADC conditioning circuit as shown in Figure 5.97. Given that the multiplexor select time is 5 μs, the

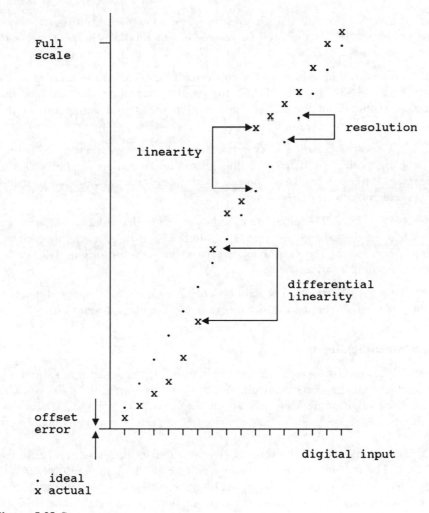

Figure 5.98 *Converter accuracy*

S/H aperture time 0.1 μs, the ADC conversion time 14.8 μs, and the buffer load time 0.1 μs, then the

conversion rate $= 1/(5 + 0.1 + 14.8 + 0.1)$ MHz $= 1/20$ MHz $= 50$ kHz

For a DAC, the conversion time is the time taken between a digital input arriving and the corresponding change in the output. It is sometimes referred to as the 'settling time' of the DAC.

5.13.4.5 Stability. In the case of DAC devices, a change in temperature of the device will cause a slight modification to the output. The exact specification is given in the data sheet supplied by the manufacturer.

5.13.4.6 Monotonicity. A converter is monotonic if, for all increases in the digital input, the analogue output increases or remains the same, but does not decrease.

Similarly, for all decreases in digital input, the analogue output decreases or remains the same, but does not increase. Non-monotonic characteristics are affected by temperature.

5.13.4.7 Linearity. The linearity of a convertor is also called the relative accuracy (Figure 5.98). For the DAC, this is the maximum deviation of the analogue output from the straight line joining the digital end points of the range.

5.13.4.8 Differential Linearity. A far better measure of linearity errors is differential linearity, which is defined as the maximum deviation in the analogue difference between two adjacent output codes from the ideal value, *i.e.*, the full range divided by the number of steps, $V_{max}/2^n$.

5.13.4.9 Offset Error. The offset error is the deviation in analogue output from zero when the digital code for zero is selected. This is in effect the amount by which the graph of the output fails to pass through the origin. It may be adjusted in compensation.

5.13.4.10 Gain or Scale Error. The gain or scale error is the percentage difference in slope between the actual transfer function and the ideal straight line.

5.13.5 Multiplexing

A multiplexor (multiplexer) is a device with multiple analogue inputs, but only one analogue output. Each analogue input, there are normally 16, is switched in turn (time-division multiplexing) to the single output. A single A/D converter can therefore be used for many analogue signals, thereby reducing cost (Figure 5.99).

Considerations include the throughput rate, the settling time, and transfer accuracy. The throughput rate is the highest rate at which the multiplexer can switch between channels. This is limited by the settling time, the time taken for the output voltage to correspond correctly to the input voltage. The transfer accuracy is the percentage I/O error.

5.14 SELECTION CRITERIA

One of the most important considerations in interfacing is the performance of the whole system. The specification of the interface board will determine how fast signals can be converted. However, this data still has to be transferred to the PC memory where it can then be stored and processed, and the results can be displayed. The transfer rate of the system as a whole is determined by the slowest factor, either the data transfer rate or the board throughput.

Data can be transferred to the memory either by the software or hardware. Software is slow but cheap, and uses polling and interrupts. Hardware is fast but expensive, and uses DMA.

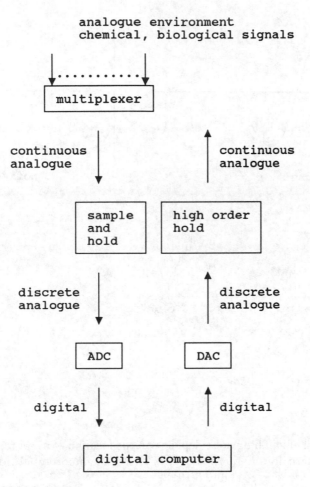

Figure 5.99 *Multiplexing*

5.15 INTERFERENCE AND NOISE

Electronic systems are subject to *noise*. Noise can be considered as any unwanted signal occurring within a system or emanating from it. Noise can be random or repetitive, occurring continuously or in isolated bursts (Figure 5.100). All signals from transducers have a noise component. A common measurement of noise is the signal to noise ratio (SNR), which is defined as the ratio of the signal power P_s to noise power P_n (SNR $= P_s/P_n$). Where the total noise is contributed from all sources, in effect,

$$\text{SNR} = P_s/P_n + P_{n2} + P_{n3} \ldots P_{nx}$$

It is usual to express the SNR in decibel units.

$$\text{SNR} = 10\log P_s/P_n$$

meter, which is listening, is called the ACCEPTOR and controls the NRFD and NADC lines. The names and lines each device controls are reversed when their respective listen and talk roles are reversed. The following sequence of events occurs:

1   REN (REmote ENable) line goes low
    All instruments on the bus go into the remote state
2   IFC (InterFace Clear) pulses low
    Stops activity on the bus
3   ATN (ATtentioN) line goes low
    Informs all instruments that the following data form an interface message
4   NRFD (Not Ready For Data) line is high. This line is one of the three handshake lines and because it is high all instruments are ready to accept data. The line stays low until the slowest instrument is ready
5   The ASCII character ? for UNL (UNListen) is put on the data lines. This bus instruction deselects any instruments that may have previously been set up as listeners. The character stays on the bus for the whole of the following handshake cycle
6   DAV (DAta Valid) line goes low
    Controller asserts data is valid
7   NRFD (Not Ready For Data) lines go low
    Instruments say, do not change data while it is being read
8   NDAC (Not Data ACcepted) line goes high when all   Handshake
    instruments have read the data   sequence
9   DAV line goes high
    Data no longer valid
10 NADC line goes low
    Instruments remove data accepted indication
11 NRFD line goes high
    Instruments ready for next data
12 Controller puts character U on the bus indicating that it is going to be the talker
    Steps 6 to 11 are repeated, resulting in the completion of the handshake
13 Controller puts character $ on the bus. This is the pH meter LISTEN address
    Steps 6 to 11 are repeated; however, only the pH meter is doing the handshaking as only it recognizes its listening address
    Until otherwise specified, steps 6 to 11 are repeated for each byte of transmitted data and are controlled by the pH meter and controller only
14 ATN (ATtentioN) line goes high
    Interface message is finished
    Device-dependent message follows
15 P on the bus. The pH meter understands this to mean pH. This character and others that follow are totally device dependent and are therefore only understood by the target device
16 R, range
17 ?, autorange

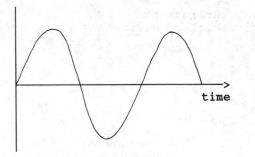

signal   noise

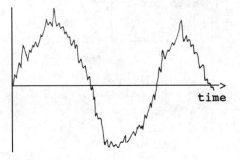

signal + noise

Figure 5.100 *Noise*

The method of handling noise depends on the location of its source; it may be generated externally or internally. It is therefore important to identify and isolate each noise source (Figure 5.101).

5.15.1 Internal Noise

It is unlikely that internal noise will be a problem. White noise contains the same amount of power across the bandwidth. There are two major mechanisms that cause it: Shot noise, also called Schottky noise, is a noise current arising from the discrete nature of electronic charge (*i.e.*, the electron); and Thermal or Johnson noise, which is due to the random movement of thermally charged carriers.

At lower frequencies, other noise sources become dominant and the noise component increases as the frequency decreases. This is called flicker or $1/f$ (where f is frequency) noise. In low frequency applications this is the most critical internal source of noise.

5.15.2 External (Environmental) Noise

This is called electromagnetic interference (EMI), but is more commonly referred to as radio frequency interference (RFI) even if it does not interfere with

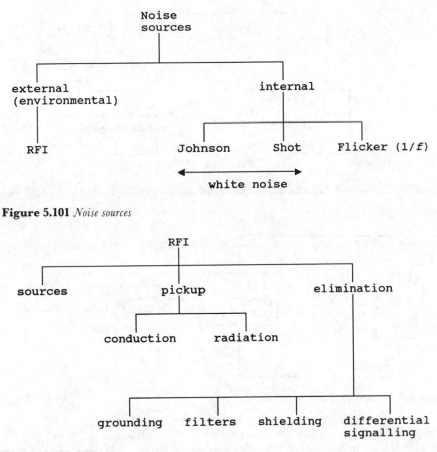

Figure 5.101 *Noise sources*

Figure 5.102 *RFI*

radio reception. Unlike internal noise, RFI can cause device malfunction and even failure. The major considerations are therefore (Figure 5.102) external noise sources, how the noise enters the system, and how the system can be made noise proof.

Statutory regulations governing the emission of EMI from electronic equipment exist in most countries, for example, the Federal Communications Commission (FCC) in the US and the Verband Deutsche Electrotechniker (VDE) in Germany. Sources of external noise were mentioned in Chapter 2. These include large electric motors and arc welding equipment.

Noise pick-up or transfer from one system to another may be by conduction or radiation (Figures 5.103 and 5.104). For conduction to occur there must be a complete circuit between the source of the noise and the receiver. Should the power supply be used by other devices, noise may be transmitted down the power lines. Separate power sources address this problem. Radiated noise is via electromagnetic radiation. By 1995, the European Community Directive 89/336/EEC will require manufacturers and importers to test and certify that products

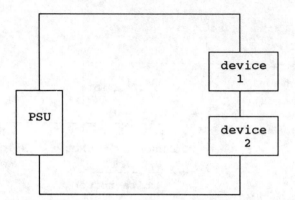

Figure 5.103 *Conducted noise*

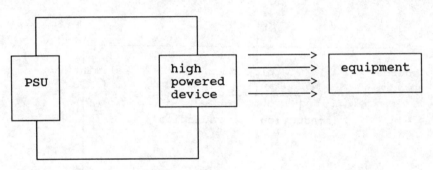

Figure 5.104 *Radiated noise*

comply with standards for electromagnetic compatibility (EMC) including radiated emissions, emissions on the mains and signal cables, transients, and electrostatic discharges, thereby providing a suitable environment for the operation of equipment.

There are various noise proofing and external noise elimination techniques depending on the source and type of noise (Figure 5.105). An ideal ground is a point that has zero resistance, is at a constant potential, and serves as a common reference point. It is important that connections to ground are clean and secure. Correct use of the 'ground' can eliminate many noise problems. A common

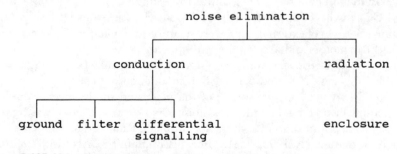

Figure 5.105 *Noise elimination*

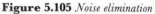

source of conducted noise is through the power lines; as mentioned in Chapter 2, the use of UPS eliminates this problem.

An electric signal is conveyed by the movement of electrons, and a return wire is needed to complete the circuit. If the return wire is grounded and the data are sent by putting an absolute voltage on the signal wire, for example + 10 V, then the transmission is said to be unbalanced. However, this arrangement is sensitive to external noise. A significant improvement is to employ differential signalling, which uses two wires for a single data path. Instead of earthing the return path and driving only one wire in an unbalanced manner, signals can be driven on both wires. For example, a TRUE signal may be + 5 V on one wire and − 5 V on the other. The FALSE signal would be the reverse of these polarities. Noise is induced in both wires without affecting the difference in voltages which represents the data signal, *e.g.*, RS 422 standard.

5.16 INSTRUMENTATION

A wide range of laboratory instrumentation is available. In the simplest form the analogue output from, for example, a spectrophotometer, may be fed into an A/D converter and then into a computer for data capture and processing. However, the trend is for chemical instruments to have a dedicated on-board microprocessor. The microprocessor controls data capture, signal processing, data processing, and data presentation—the term 'intelligent instrumentation' is sometimes used. As such, these instruments have an optional communication port. Of particular note is the Hewlett-Packard 8452A diode array UV/VIS spectrophotometer, which can only be controlled through the IEEE 488 interface. In addition to the IEEE 488 port, there are two programmable ports, one 8 bit input and one 8 bit output, that can be used to interface with other equipment such as a peristaltic pump. The instrument has buffered I/O. The I/O buffers are 256 bytes long and the 6588 byte data buffer is capable of storing one full spectrum. The instrument may be controlled by means of high-level language statements. Hewlett-Packard provide a suite of prewritten programs for data collection and manipulation.

By contrast the Perkin Elmer LS-5 spectrofluorimeter uses an RS-232-C interface for communications. As such it has a switchable baud rate (300, 600, 1200, and 2400), full duplex capability, 7 bit word length with parity, and a user panel for control.

Automatic balances typically have either RS-232-C or 20 mA current loop interfaces allowing automatic computer control.

5.17 CHROMATOGRAPHY BOARDS FOR PCS

The microprocessor-based 'intelligence' may reside as we have seen in the instrument or in the associated computer. A cost-effective solution to liquid and gas chromatography (LC and GC) data collection is the use of microprocessor-controlled data-acquisition cards (p.c.b.s) that plug into a PC. An example of this is the Data Translation DT2802 24 bit chromatography board for PC AT compatible computers. We will now examine the specification for such devices to illustrate the importance of many of the technical considerations raised earlier.

Input channel types: single ended (SE) or differential (DI)
Normal mode rejection ratio (NMRR): 60 dB
Common mode rejection ratio (CMRR): 72 dB
Connection: RS/232;
            full duplex; one start bit, one stop bit; eight data bits; no parity; 300, 1200, 2400, 4800, 9600, 19.2K, 38.4K baud rate (software selectable)
Data capture: simultaneous sample and hold
Resolution: programmable for 18 through 24 bits
Sampling rate: 100 Hz at 18 bits
                2 Hz at 24 bits
Differential
    Nonlinearity: 1 LSB max
    Monotonicity: guaranteed
Interface: IBM PC AT bus (8 bit data path) with I/O mapping

The number of input channels determines the number of devices that can be connected to the data acquisition card; typically, this varies from 4 to 16, but it can be considerably more. Input channels can be configured to either single ended or differential. Given 16 input channels, the card may be set to 16 SAE or 8 DI. As we have seen, the use of two wires for differential signalling provides considerable improvement in noise immunity. This is sometimes called common-mode rejection (CMR) the performance of which is measured by the common mode rejection ratio (CMRR). This can be contrasted with the NMRR. The decibel notation, as we have seen, is a convenient method of expressing ratios of power (and amplitudes) that extend over large ranges. They are easily remembered for certain values such as powers of 10; recall that $\log_{10} n$ is simply the

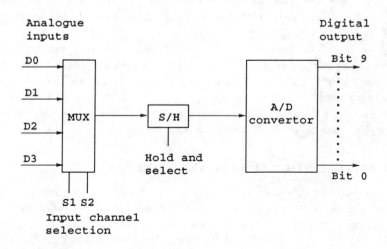

Figure 5.106 *Single sample and hold*

number to which 10 must be raised to equal n. Gains of 10, 100, and 1000 are simply 10 dB, 20 dB, and 30 dB, respectively. Similarly, attenuations of 1/10, 1/100, and 1/1000 are simply -10 dB, -20 dB and -30 dB. The other values of significance are 3 dB, and -3 dB which represent a doubling or halving, respectively. In the previous example, the change from NMRR to CMRR represents a 12 dB improvement in the signal quality.

The connection to the card is via the RS 232 port and is programmable to a variety of configurations.

Single S/H circuits Figure 5.107, while allowing for many input channels by means of a multiplexing circuit (MUX) only 'freeze' the status of one data signal at a time. The MUX connects the multiple inputs to a single output. At any time, one of the inputs is selected to be passed to the output. In this example, it is a 4 to 1 multiplexor. To select one of the four possible inputs, a 2 bit selection code is needed. This is implemented by means of the two select lines, S1 and S2.

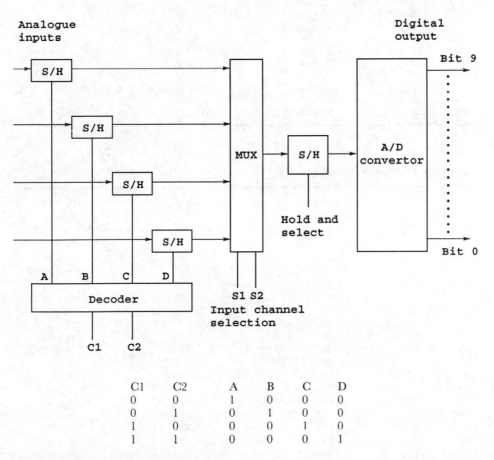

C1	C2	A	B	C	D
0	0	1	0	0	0
0	1	0	1	0	0
1	0	0	0	1	0
1	1	0	0	0	1

Figure 5.107 *Simultaneous sample and hold*

S1	S2	output
0	0	D0
0	1	D1
1	0	D2
1	1	D3

Simultaneous S/H circuits have a separate S/H circuit on each input channel. The decoder, which we first met in memory circuits, has a number of output lines only one of which is asserted at any time depending on the input value. In this example, the decoder has two input lines and four output lines (2^2) thereby reducing the number of input lines.

Table 5.2 *Combination table*

No. of bits	No. of combinations	No. of bits	No. of combinations	No. of bits	No. of combinations	No. of bits	No. of combinations
1	2	11	2K	21	2M	31	2G
2	4	12	4K	22	4M	32	4G
3	8	13	8K	23	8M	33	8G
4	16	14	16K	24	16M	34	16G
5	32	15	32K	25	32M	35	32G
6	64	16	64K	26	64M	36	64G
7	128	17	128K	27	128M	37	128G
8	256	18	256K	28	256M	38	256G
9	512	19	512K	29	512M	39	512G
10	1024 (=1K)	20	1024K (=1M)	30	1024M (=1G)	40	1024G (=1T)

Table 5.3 *Resolution table*

Resolution bits	Number of steps	Step size range/%	10 V range
8	256	0.39	39 mV
9	512	0.195	19.6 mV
10	1024 (= 1K)	0.098	9.8 mV
11	2K	0.049	4.9 mV
12	4K	0.024	2.4 mV
13	8K	0.012	1.2 mV
14	16K	0.006	0.6 mV
15	32K	0.003	0.3 mV
16	64K	0.0005	0.15 mV
17	128K	0.00075	75 μV
18	256K	0.000375	37 μV
19	512K	0.0001875	18 μV
20	1024K (= 1M)	0.0000937	9.0 μV
21	2M	0.0000468	4.6 μV
22	4M	0.0000234	2.3 μV
23	8M	0.0000117	1.1 μV
24	16M	0.0000058	0.58 μV

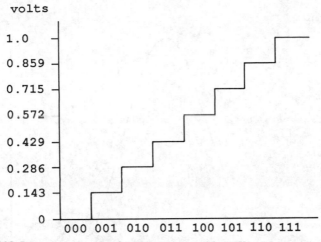

Figure 5.108 *Dynamic range value for 3 bit converter with a 1 V range*

We can use the binary combination table that we first met in Chapter 1 to help us calculate the resolution of the converter (Tables 5.2 and 5.3).

Strictly speaking, the dynamic range value, which is the number of steps less one (for zero), should be used; however, given the size of the numbers this is often ignored. It is considered that for low noise chromatographic purposes, the A/D converter should have a resolution greater than or equal to 20 bits, *i.e.*, of about 1 μV. High-resolution devices are sensitive to noise and temperature drift. Various design features are incorporated to minimize the effects of noise and drift. These include circuits shielded in a steel enclosure, shielded electrical cables, the use of common-mode rejection, and autocalibration.

For a 3 bit converter with a 1 V range, the

$$\text{number of steps} = 2^3 = 8 \text{ (Figure 5.108)}$$

and the

$$\text{resolution} = \text{voltage range}/(2^3 - 1)$$

which for large values approximate to

$$\text{voltage range}/(2^3 - 1) = 1/7 = 0.143 \text{ mV}$$

Figure 5.109 shows the peaks (A and C), which you do not see, that cause the problems if you were to use this 3 bit converter.

In high noise chromatographic applications such as high-performance liquid chromatography (HPLC) there can be 5 μV noise on a full-scale signal of 1 V. In such cases a lower resolution detector could be used satisfactorily.

The throughput varies inversely with resolution from 100 Hz at 18 bits to 2 Hz at 24 bits. As we have seen, according to the Nyquist sampling criterion, a data-acquisition system must sample the input signal at least twice as fast as the highest frequency component.

The card interfaces to the IBM PC AT bus (8 bit data path) with I/O mapping.

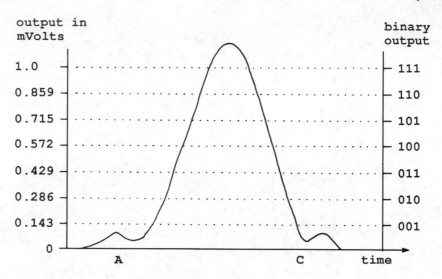

Figure 5.109 *Hidden peaks with 3 bit converter*

5.17.1 IBM PC/XT/AT Data Transfer

5.17.1.1 Programmed I/O. With the requirement for only low throughput ($\leqslant 40$ kHz), programmed I/O data-acquisition boards may be used. As we have previously seen, this is a simple but inefficient method of data transfer. For each data item to be transferred, a small program must be executed, which takes time. Furthermore, there is direct CPU involvement in each data transfer and the data-acquisition process must therefore compete with other devices demanding CPU time.

5.17.1.2 Polling. Programmed I/O (PIO) boards that do not have the interrupt facility require software on the computer to poll the board continually to control data transfer. This is an inefficient method of data transfer with the polling rate affecting the throughput. Only one data word is transferred at a time.

5.17.1.3 Interrupts. PIO boards that support the interrupt facility allow for data transfer 'on demand'. Only one data word is transferred at a time.

5.17.1.4 Direct Memory Access. Using DMA permits a significant improvement in both the speed of data transfer (100 kHz) and the number of data words (32K). As we have seen there is no CPU involvement. There are single- and multi-channel DMA cards available for PCs. The length of the data word transferred depends on the bus standard. The XT is an 8 bit standard and the AT is 16 bit.

PC and PC XT systems: 8086 CPU (8 bit ISA)
                                      8088 CPU (8 bit ISA)
PC AT systems: 80286 CPU (16 bit ISA)
                         80386 CPU (16 bit (ISA)
PS/2 MC Bus: 80286 CPU (16 bit MCA)
                     80386 CPU (32 bit MCA)
                     80486 CPU (32 bit MCA)

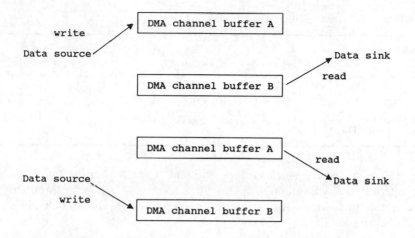

Figure 5.110 *Flip-flop buffer*

EIS Bus: 80386 CPU (32 bit EISA)

Single- or multi-channel DMA cards are available. As the name implies, single-channel DMA cards provide a single data channel. In a dual-channel DMA card, there are two data channels with buffers. When one buffer is full, for example, DMA buffer B, the data will be read from it; however, data acquisition still occurs in the second buffer A. When buffer B is empty and A is full, the read and writer operations are swapped over or 'flipped'. This approach is sometimes referred to as flip-flop buffering; it allows continuous data acquisition without interruption (Figure 5.110). With fast, dual-channel DMA, the limiting factor for the data-transfer rate becomes the performance of the hard disk (Table 5.4).

Such PC cards must have the associated software to control them. Simple to use, menu-driven software is available that includes a library of subroutines and device drivers. The subroutines and device drivers allow predetermined, specific tasks to be performed, such as controlling the specific hardware cards being used. Tasks include setting the baud rate, interrupt masking and enabling, resetting the system, and configuring the I/O ports.

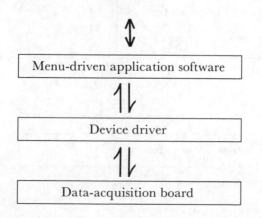

Table 5.4 *Summary g transfer parameters*

Transfer method	Bus standard	Typical throughput/kHz
Polling	XT/AT	20
Interrupt	XT/AT	40
Single-channel DMA	XT/AT	100
Dual-channel DMA	AT	250

The use of microprocessor-based, 'intelligent' data-acquisition cards such as these permit data acquisition to operate in 'background' mode. In this mode, the card is able to operate with minimum intervention from the host PC; in effect, this is multitasking.

5.18 ANALOGUE TO DIGITAL VALIDATION

As we will see in Chapter 9, computer systems must be validated. In this context A/D interfaces must be validated for accuracy, linearity, reproducibility, *etc.* The performance of electronic devices degrades with time and therefore such devices require not only validation but also periodic revalidation according to a set plan.

Any such validation requires an extremely accurate analogue test value or set of values that can be traced to a primary standard. Microprocessor modules are available to assist interface validation. Such devices hold a test pattern, in an EPROM chip, which is calibrated to a standard that can be traced to a primary standard. This test pattern is converted to an analogue signal, which, under microprocessor control, is input into the device to be validated at a variety of voltages and rates (Figure 5.111).

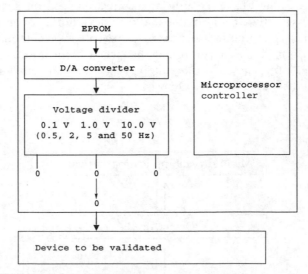

Figure 5.111 *Analogue to digital validation*

CHAPTER 6

Networking

With the increasing availability of PCs and the introduction of standards for communication between computers, there has been a progressive trend towards distributed data processing (DDP).

A *computer network* is an interconnection of devices and computers capable of information sharing. This can be contrasted with a *distributed system*, which is a special type of network with a high degree of cohesion and transparency. A distributed processing system is one in which several autonomous processors interact in order to achieve a common goal. The processors co-ordinate their activities and exchange information by means of a communication network.

The advantages of networking include the following: incremental cost (scaling or extensibility); the end of geographical tyranny, 'soft' or graceful performance degradation; local control; and shared use. The number of devices and computers on the network can be increased according to user needs. Furthermore, the extent of the network can also be increased as appropriate. Previously remote devices and computers can be linked to others and, whereas with centralized computing, all users will be affected should the main computer fail, with networked computers, local failure does not prevent other computers and devices from continuing to work. Each user has their own computer and therefore local control. This can be contrasted with a central computer with centralized data and resource control. Expensive software and hardware may be shared and is available to all network users. This progressive change to networking is at the expense of mainframe computers and is called 'downsizing'.

There are three main types of DDP environment, heterogeneous, network, and homogeneous. In a heterogeneous environment, each computer has its own operating system, and communication to other machines is by explicit reference. A network is an arrangement of application machines, typically single-user workstations, and one or more 'server' machines. The server machines provide network-wide services such as file management and are therefore also called 'file servers'. The users are aware that communication to other resources is via a network. A homogeneous sytem is a common operating system shared by a network of computers in which the user is unaware of the distributed nature of the resources.

6.1 COMMUNICATION LAYERS

We considered data transfer when we looked at interfacing. In the case of the RS 232 link we met a simple protocol to establish a communications link, and to co-ordinate the timing for data transfer and closing or clearing the line. However, the reliable transmission of data between equipment over a network is more complex than this. Consider making a telephone call to a scientist who works in another laboratory in a foreign country. There are many complex functions that must be performed. Most of this complexity is hidden from the users. This complexity can nevertheless be controlled by the logical decomposition of the system into subsystems or layers. There are clearly defined layers with associated tasks that must be performed.

After the successful connection of the call, the scientists may exchange data on the *application* under consideration, *e.g.*, chromatographic results. We can identify this as the *application layer*.

With international telephone conversations it is necessary to resolve language problems and *present* the information in a language that both parties understand. For example, both parties may not be able to speak each other's language but they both may understand Danish. This can be called the *presentation layer*.

The caller must set up a *session* by asking to be connected to one person in particular. If the named scientist is called away temporarily the same session can be continued later (re-established). As both parties cannot talk at the same time, flow control is needed and dialogue synchronization must be provided to set up and clear a communication channel, *i.e.*, there are rules of conversation that indicate when a conversation is to begin, when one is to speak and listen, and when the conversation is over. This is called the *session layer*.

When the call has been established it is important to ensure that the messages are *transported* without loss. Should communication be difficult, both parties can agree to hang up and redial. After the conversation has been completed both parties hang up. This is called the *transport layer*.

The telephone system is a complex *network* of many subscribers on an international scale. There must be a mechanism for dialling the number of the laboratory the caller wishes to contact. Should it be a wrong number the caller is informed accordingly. This is the *network layer*.

Any words that are not clearly received must be notified to the other scientist and the appropriate action taken, *i.e.*, retransmission. There are previously agreed words to implement this protocol. This is the *data-link layer*.

The *physical* wires and electrical interface provide the medium by which the telephone conversation or bit stream may be transmitted. This is the *physical layer*.

6.2 COMMUNICATION ADDRESSING

Let us now use another analogy, sending a letter. The contents (data) are placed in an envelope and the destination address is written on the front. The address starts with the number, then continues with the street, and then the city (Figure 6.1). There is a hierarchy to the address.

All the letters posted are collected and sorted. In the sorting process the name of the city is the first address item used. For example, all letters to Copenhagen are collected together. When this part of the address has been read for sorting it is

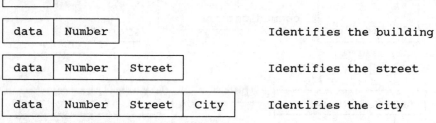

Figure 6.1 *Data addressing*

no longer of use (and could in fact be discarded). The Copenhagen letters are then subdivided into all those going to, for example, Vesterbrogade. Again, when the name of the street has been used for sorting it is no longer of use. Finally, the number or name identifies the actual building, for example Tivoli. The delivery process uses the address in the reverse order in which it was written.

6.3 OPEN SYSTEMS INTERCONNECTION MODEL

Different computers have different operating systems and indeed different ways of representing data. The ASCII code is not universally used! Historically there were 'closed' computer communities in which each computer manufacturer had their own standards for computer communications (Figure 6.2). In the 1970s, the ISO developed the reference model for OSI in order to address the problem of universal interconnectivity. This allowed: 'Any application in any computer that supports the appropriate standards to freely communicate with an application in any other computer supporting the same standard irrespective of its origin of manufacture'. The important consideration here is the establishment of manufacturer-independent standards. Examples of application processes that may wish to communicate in an open way include desktop computers and intelligent instrumentation.

Communications software is complex and must address many issues and problems. As an aid to this complexity the OSI standard is divided into layers each with a well-defined function operating to defined protocols with defined

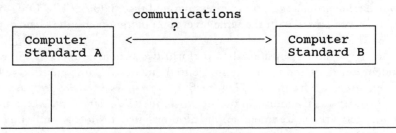

Figure 6.2 *Communications link*

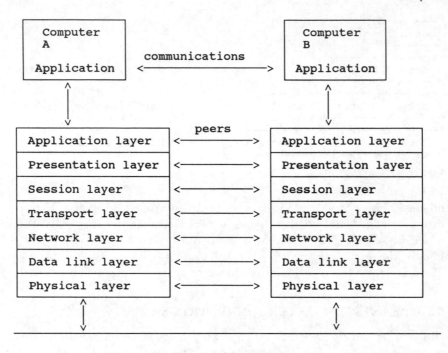

Figure 6.3 *The OSI model*

interfaces. In particular, we note the logical decomposition of complex systems into subsystems (layers), standard interfaces between layers, and symmetrical functions, *i.e.*, peer stations perform the same functions.

In the case of data transfer between computers there are analogies with the telephone link. We note that for a reliable communications link there must be a high degree of co-operation between the two scientists and computers. Further, it is possible to control the complexity by modularizing the tasks into layers that are mutually dependent; this is referred to as the *communications architecture*. The advantage of this is that it is possible to modify correctly-designed modules without having to change any of the surrounding modules. The OSI model consists of seven layers, with the same layers on different computers being called *peer layers* (Figure 6.3).

These layers can be considered to perform two primary functions, network-dependent functions and application-oriented functions. Consequently, we have three operating environments (Figure 6.4): the *network environment*, which is concerned with the communications network; the *OSI environment*, which builds upon the network-environment application-oriented protocols and standards that allow users to communicate in an open manner; and the *real-systems environment*, which builds upon the OSI environment and is concerned with the manufacturers' own software.

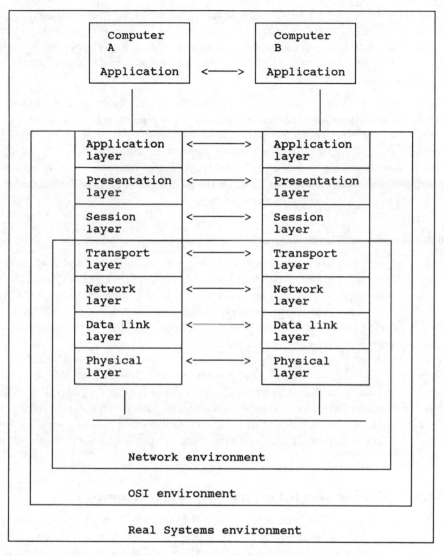

Figure 6.4 *OSI environments*

6.3.1 Network Dependent Layers

The lowest three layers are network dependent—their detailed operation will vary depending on the type of network.

6.3.1.1 Physical Layer. The physical layer is responsible for both the physical and electrical interface and provides the means for transmitting the pulse train. Layer functions include physical transmission of the data stream through the medium, collision detection for carrier sense multiple access with collision detection (CSMA/CD) (considered later), and providing appropriate interface signal levels, data rates, and control signals.

6.3.1.2 Data-link Layer. The data-link layer builds upon the lower physical layer in order to ensure reliable data transfer. As such it provides error-protected transmission and access to the communications medium. Layer functions include establishing and releasing the link, data framing and synchronization, sequence control detection and correction of transmission errors, and flow control.

6.3.1.3 Network Layer. The network layer is responsible for establishing the connections between selected devices across what is possibly a large and complex network using error-protected paths provided by the data-link layer. It also defines how several network connections may share a link by determining the appropriate switching and routing of messages. Layer functions include routing (address location), call set-up and clearing, selection of primary-secondary routes, and network connection multiplexing.

6.3.2 Transport Layer

The transport layer acts as the interface between the higher application-oriented protocols and the network-dependent layers. As such it provides the session layer with a reliable data-exchange capability that is independent of the underlying network type. The transport layer is responsible for establishing, controlling and releasing end to end connections. Layer functions include end to end message transfer, connection management, error control, and flow control.

Due to the different type of network, the transport layer offers a number of *classes of service* with varying *quality of service (QOS)* provided by the different network types. There are five classes, class 0, which provides only the basic functions needed for connection establishment and data transfer, class 1, which is for basic error recovery, class 2, which is for multiplexing, class 3, which is for error recovery and multiplexing, and class 4, which provides full error- and flow control.

6.3.3 Application-oriented Layers—the OSI Environment

6.3.3.1 Session Layer. The session layer establishes and terminates the communication period (session or dialogue) in an orderly manner thereby synchronizing data exchange. Session control includes the control of half-duplex communication. Layer functions include the following: setting up (and clearing) a communication channel between two correspondent layer protocol entities for the duration of the transaction; interaction management, also called dialogue control (data exchange may be duplex or half duplex; half duplex must be synchronization; and recovery without data loss from communication problems during a session).

6.3.3.2 Presentation Layer. Different computers may represent data in different ways. It is necessary to select the appropriate syntax. The presentation layer modifies the data in the application to one suitable for the communications system. The presentation protocol defines rules on how the data will be exchanged into a common neutral language, *e.g.*, from EDCDIC to ASCII.

Layer functions include performing data exchange, data encryption and de-cryption if needed, and data compression and decompression if needed.

6.3.3.3 Application Layer. This layer is concerned with the application rather than with the network itself. The application layer allows the computer's operating system to use the system as a local device. The detailed implementation and operation of the communication subsystem is transparent to the user. The application layer provides the user interface to a range of network-wide (distributed) services including identification of the communication partners and establishing availability, authorization and validity checks, cost allocation, and database queries and entries.

Amongst the services available, are those described here. File Transfer, Access and Management (FTAM), for instance, provides access to a remote file system. This allows the user at one end of a link to access and manipulate files at the other end. Remote file manipulation (creation, amendment, deletion, *etc.*) are typical FTAM tasks. Job Transfer and Manipulation (JTM) provides the facility for submitting a job to a remote station for processing and subsequently monitoring its execution. Virtual Terminal (VT) allows standard communication using a local terminal, *i.e.*, manufacturer-independent communication. Message Hand-ling Service (MHS) provides a general facility for two stations to exchange messages. Manufacturing Message Service (MMS) provides the facility for communication to intelligent instrumentation and controllers. Lastly, Directory Services (DS) allows the use of symbolic names for addressing.

The structure of the seven-layer OSI model is shown in Figure 6.5.

6.3.4 OSI in Operation

Let us now examine the principles of operation of the OSI model using the letter analogy. The application layer generates data that can be sent to another computer. For complete communication, every application needs a unique address as there may be several applications running at a given time. These addresses are called *service access points* (SAPs) (Figure 6.6). Further, each com-puter must have a unique address. The combination of data from the higher level and the controlling address information is called a *protocol data unit* (PDU). The PDU is identified as the presentation PDU (PPDU) in the presentation layer, the session PDU (SPDU) in the session layer, *etc.* The data to be transmitted are progressively combined with additional addressing information. This package is then put onto the network (Figure 6.7). The information in each PDU header is used by the peer layer in the destination computer for addressing. When delivering this package, the address is read in the reverse order in which it was written and then the data are delivered to the user as shown in Figure 6.8.

6.4 TRANSMISSION CONTROL PROTOCOL/INTERNET PROTOCOL

Vendor independent Transmission Control Protocol/Internet Protocol (TCP/IP) has been used since about 1975, mainly on defence-data networks in

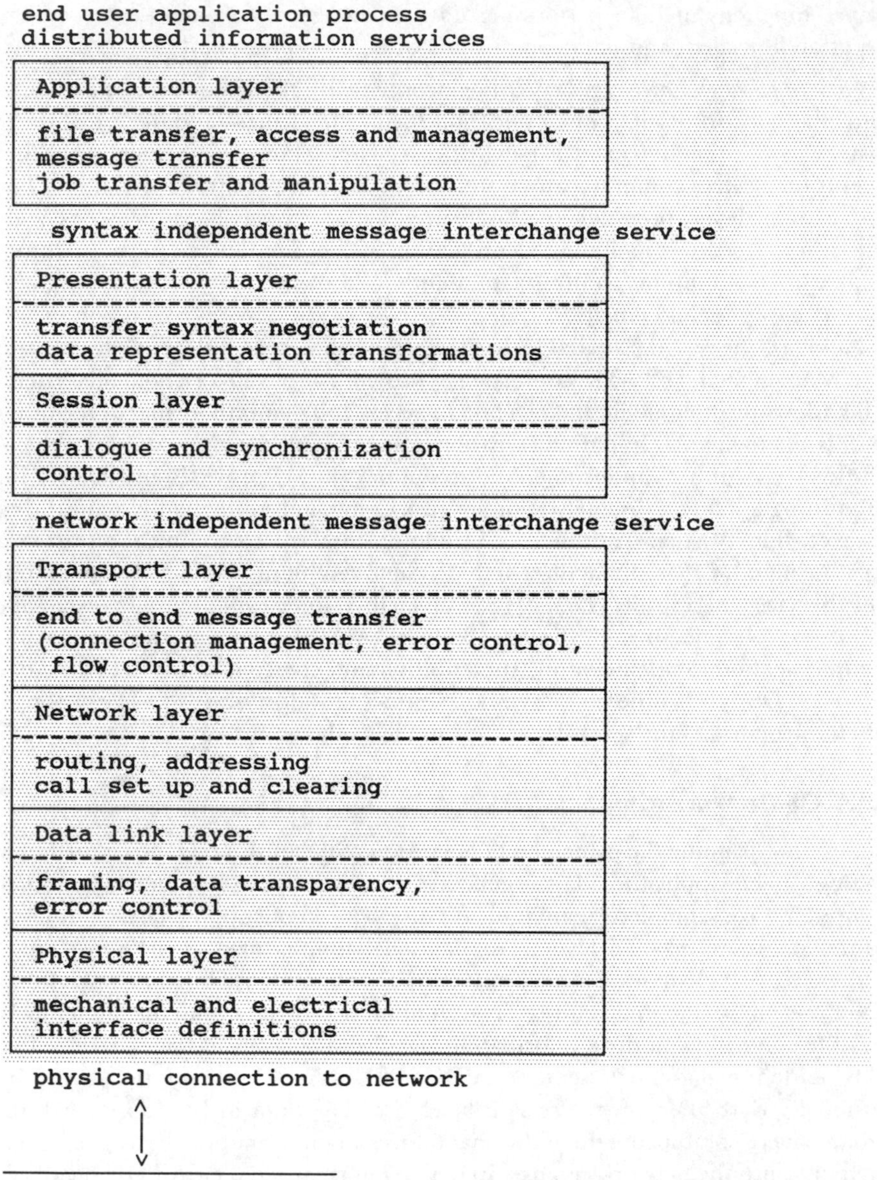

end user application process
distributed information services

```
┌─────────────────────────────────────────┐
│ Application layer                         │
│ - - - - - - - - - - - - - - - - - - - - - │
│ file transfer, access and management,     │
│ message transfer                          │
│ job transfer and manipulation             │
└─────────────────────────────────────────┘
```

syntax independent message interchange service

```
┌─────────────────────────────────────────┐
│ Presentation layer                        │
│ - - - - - - - - - - - - - - - - - - - - - │
│ transfer syntax negotiation               │
│ data representation transformations       │
├─────────────────────────────────────────┤
│ Session layer                             │
│ - - - - - - - - - - - - - - - - - - - - - │
│ dialogue and synchronization              │
│ control                                   │
└─────────────────────────────────────────┘
```

network independent message interchange service

```
┌─────────────────────────────────────────┐
│ Transport layer                           │
│ - - - - - - - - - - - - - - - - - - - - - │
│ end to end message transfer               │
│ (connection management, error control,    │
│  flow control)                            │
├─────────────────────────────────────────┤
│ Network layer                             │
│ - - - - - - - - - - - - - - - - - - - - - │
│ routing, addressing                       │
│ call set up and clearing                  │
├─────────────────────────────────────────┤
│ Data link layer                           │
│ - - - - - - - - - - - - - - - - - - - - - │
│ framing, data transparency,               │
│ error control                             │
├─────────────────────────────────────────┤
│ Physical layer                            │
│ - - - - - - - - - - - - - - - - - - - - - │
│ mechanical and electrical                 │
│ interface definitions                     │
└─────────────────────────────────────────┘
```

physical connection to network

Figure 6.5 *OSI layer functions*

the US, but it is also supported by many computer manufacturers. The network layer is called the IP and the transport layer is called the TCP. This protocol has been incorporated into some versions of UNIX and is a popular implementation of the LAN. It can be considered as a layered model and has some similarities with the OSI model (Table 6.1).

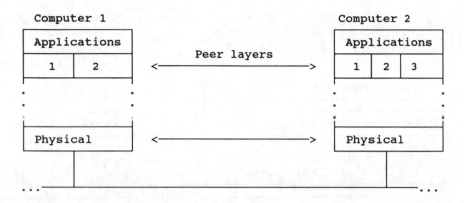

Figure 6.6 *OSI communications*

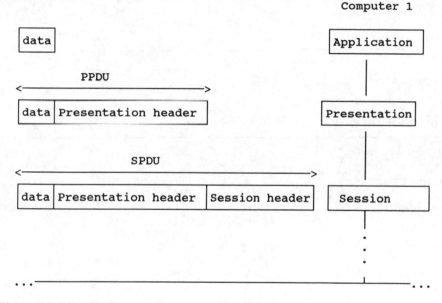

Figure 6.7 *OSI addressing*

6.5 NETWORKS

With a network, all the computers are autonomous and there is no central controlling device with subordinate computers. All types of equipment can have access to the network: workstations, intelligent instrumentation, process plant, *etc*. In general, these types of equipment are called data terminal equipment (DTE). Two categories of network, LANs and wide area networks (WANs), can be distinguished as shown in Table 6.2.

Local Area Networks consist of interconnected computer based DTE distributed within either a building or a group of buildings. As such they are owned, controlled, and maintained by the user organization; they are less commonly

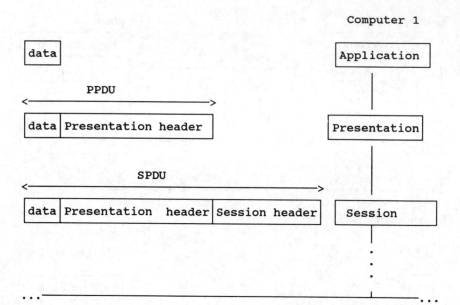

Figure 6.8 *OSI data delivery*

Table 6.1   *Comparison of TCP/IP and OSI layers*

OSI layer	TCP/IP layer
Application	File Transfer Protocol
	Remote job entry
Presentation	
	Telnet
	Virtual Terminal Protocol
Session	
Transport	Transmission Control Protocol
Network	Internet Protocol
Data link	Network dependent
	(*e.g.*, Ethernet)
Physical	Network dependent
	(*e.g.*, Ethernet)

Table 6.2   *Local area and wide area networks*

Network	Diameter	Owner or controller	Data rate/Mbits s^{-1}
Local area	less than a few kilometres	one organization	> 1
Wide area	unrestricted	common carrier with many organizations	< 1

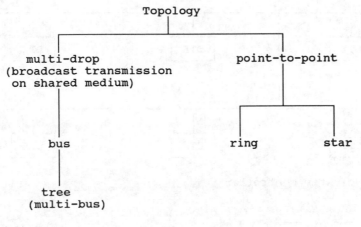

Figure 6.9 *Topology*

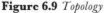

Figure 6.10 *Multidrop, bus topology*

referred to as *private data networks*. The localized nature of the network permits relatively high data transmission rates. In this book we will only consider LANs. In this context let us examine some of the characteristics of LANs, namely topology, transmission media, multiplexing, and medium-access control.

6.5.1 Topology

Topology is the geometry of the shape of the network. The two main classes of topology provide *multidrop* and *point-to-point* connections (Figure 6.9). The multi-drop topology has a single network cable that is shared by all the users. Users must *broadcast* the data onto the bus. This can be contrasted with point to point topology where there is transmission between defined points, which have *sole* use of the transmission medium. Examples of multidrop topologies are *bus* and *tree* configurations (Figures 6.10 and 6.11). Bus configurations are linear structures on which the DTE is connected. Bus topologies can be extended into an interconnected set of buses that is sometimes referred to as a tree structure (some say it resembles an uprooted tree!). Examples of point-to-point topologies are *ring* and *star* configurations (Figure 6.12 and 6.13). With the ring configuration, the cable connects the computer systems in the form of a ring or loop. There is a direct point-to-point link between neighbouring equipment, which is uni-

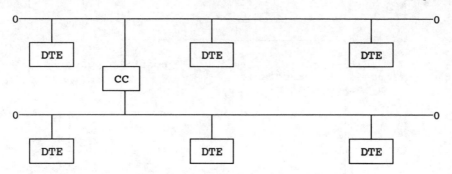

CC - cluster controller

Figure 6.11 *Multidrop, tree topology*

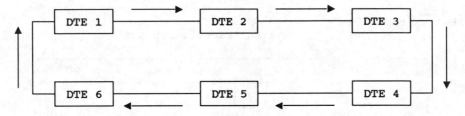

Figure 6.12 *Point-to-point, ring topology*

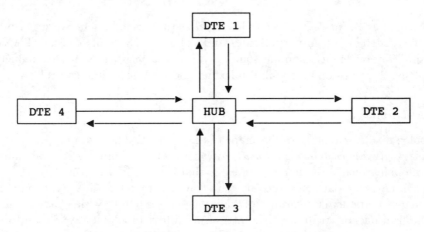

Figure 6.13 *Point-to-point, star topology*

directional in operation. Terminal-based networks are characterized by having a single central computer to which all the terminals in the network require access. As such, they have a point-to-point topology with a star configuration. Local area networks can have both point-to-point and multidrop topologies (Figure 6.14).

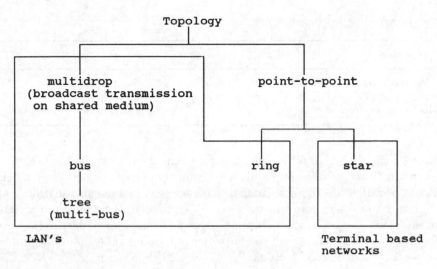

Figure 6.14 *LAN's and terminal based network topologies*

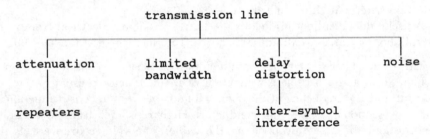

Figure 6.15 *Transmission line limitations*

6.5.2 Transmission Media

In Chapter 5 mention was made of the distances between devices. In practice, the quality of the transmitted signals is reduced by attenuation and distortion due to the effects of the transmission medium, which in turn depend on transmission speed and distance (Figure 6.15). All these errors have been quantified and form the basis of standards. The effects include attenuation, bandwidth limitations, delay distortion, and noise. As the signal moves along the transmission medium its amplitude decreases (attenuation). This can be compensated for by using amplifiers.

A digital signal (square wave) consists of a very wide range of frequencies. The bandwidth of the medium is the range of frequencies that may be transmitted. The bandwidth of the medium may be considerably less than the bandwidth of the signal causing signal distortion.

The propagation rate along a transmission line depends on the frequency. The digital signal consists of a range of frequencies. These various frequencies will therefore be subjected to different delays and so cause delay distortion.

Even with no signal on a transmission line, there will be random perturbations

Table 6.3 *Cable selection criteria*

Specification	Cable type
Low data rates	Twisted pair
High data rates	Coaxial cable
High noise environments	Shielded cable or fibre optic

called noise. This becomes a significant problem when the signal has become so attenuated that the noise amplitude is a significant percentage of the signal. This would be rather like trying to listen to someone next to a pneumatic drill.

Data transmission between two pieces of equipment is by means of a transmission medium, *i.e.*, a cable. The type, size, and installation of the cabling are of fundamental importance to the performance of the communications link and the network. The main considerations for specifying cabling include the application requirements, environmental considerations, shielding requirements, cable resistance and capacitance, and cable connections and terminations. Cable selection criteria are listed in Table 6.3.

A useful document on this field is the IEEE 'Guide on Electrical Noise: The Installation of Cables to Minimise Electrical Interference—IEEE 518—1982'.

6.5.2.1 Cables. Cables are copper conductors with plastic insulation. The signal voltage drop across a cable is a function of the conductor resistance, the line current and the frequency of the signal. The cable resistance is a function of the cross-sectional area of the conductor. High resistance leads to excessive heating and to voltage drops along the cable. The voltage drop affects the signal-to-noise ratio at the receiving end. The thicker the cable, the lower the resistance. For d.c. and low-frequency signals the dominant consideration is the resistance of the conductor. With increasing frequency (data transfer rates) other cable characteristics become important, namely cable capacitance and inductance.

In laboratories, there are four main types of transmission medium. Two-wire, open lines provide the simplest and cheapest form of connection. These are limited to short distances (up to 50 m) and low transmission rates (20 kbps) owing to susceptibility to noise pick-up. Capacitative coupling leads to cross-talk between the cables.

Twisted pair lines give better noise immunity due to the physical proximity of the wires. Twisted pair cables are made from two identical insulated copper conductors twisted together along the entire length for a specified number of twists m^{-1}, for example, 40 twists m^{-1}. The close proximity of the wires means that any electromagnetic or electrostatically induced effects affect both wires equally with the result that the relative value of the signal is not substantially degraded. The use of an earth screen around the cable further serves to reduce noise. Heavier cables are available for longer distances. Hence a bit rate of 1 Mbps per second over 100 m is possible. The distance may be increased by increasing the bit rate. Newer types of twisted pair cabling are suitable for

10 Mbps. Twisted pair cabling can be shielded twisted pair (STP) or unshielded twisted pair (UTP). For full duplex communication using differential transmission, it is possible to use two sets of screened twisted pairs contained in one cable with screening. However, at higher frequencies there are significant radiation losses due to the skin effect

Coaxial cable contains a central conductor that is shielded from external interference. This allows transmission rates of 10 to 20 Mbps over hundreds of metres. The high bandwidth of coaxial cables can be utilized in the following two ways: in baseband mode in which all the bandwidth is used for a single, high bit rate transmission path; and in broadband mode in which the bandwidth is divided into several lower bandwidth subchannels.

Coaxial cables consist of a central conductor surrounded by an insulating cylinder and a conducting cylinder on the same axis, hence the name. On top of this is a protective PVC sheath. The separation of the two conductors by the insulating cylinder affects the capacitance of the cable and allows much higher transmission rates. Coaxial cables are, however, more expensive and more difficult to connect and splice than twisted pair cables.

Optical fibres differ from other transmission media in that the carrier is a beam of light in a glass fibre, rather than an electrical signal in a piece of wire. Due to the properties of light waves a much higher bandwidth is possible, allowing very high bit rates (hundreds of Mbps). There is also the additional advantage of immunity from EMI. Optical fibres are designed for digital signals. The cables, although no more expensive than coaxial cables, suffer from complex termination and joining problems with the result that fibre optic cabling is the more expensive. The fibres are identified by the type of path, called a mode, that the light follows inside the fibre. In multimode fibres, the light takes many paths between the two fibre ends because of sidewall reflections. The result is that some paths are longer than others and produce distortion. This in turn produces pulses that overlap and therefore limit the maximum bandwidth. Monomode fibres or single mode fibres are more expensive. The light travels in a single path or mode with no reflections thereby eliminating overlap and allowing a very high data transmission rate.

The advantages of fibre optic cables are the following: no electrical interference or noise; low error rates; high data transfer rates; secure from unauthorized access by tapping; low cable weight; and no cross-talk. The following are disadvantages of fibre optic cables: they make for expensive transmitting and receiving equipment; and they are difficult to form into 'T' junctions. (They are mainly used for point-to-point communications.)

Media characteristics are summarized in Table 6.4.

6.5.3 Multiplexing

With point-to-point configurations the communications medium transmits between two points. However with multidrop configurations the medium must be shared; this is *multiplexing*. The medium bandwidth can be multiplexed in two ways, by time or by frequency (Figure 6.16).

Table 6.4 *Medium characteristics*

Medium	Distance	Bit rate
2 Wire open	10 m (< 50 m)	kbps (19.2 kbps)
Twisted pair	100 m (100 m)	Mbps (< 1 Mbps)
Coaxial	$N \times 100$ m	Mbps (< 100 Mbps)
Fibre optic	km	1000 Mbps

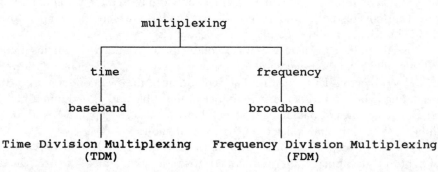

Figure 6.16 *Multiplexing*

6.5.3.1 Baseband. In baseband mode, all the available bandwidth of the cable is used exclusively for a single transmission path (channel) for a period of time; this is time division multiplexing (TDM) (Figure 6.17). This mode of operation will only allow half-duplex communication. There are two implementations of TDM, synchronous, in which each user has channel access at defined (synchronized) time intervals, and asynchronous, in which channel access is demand driven and users have random access to the channel.

6.5.3.2 Broadband. In broadband mode, the available bandwidth is divided into a number of lower bandwidth subchannels on the single cable; this is frequency division multiplexing (FDM) (Figure 6.18). This is only possible with transmission media that have a sufficiently wide bandwidth (coaxial and fibre optic).

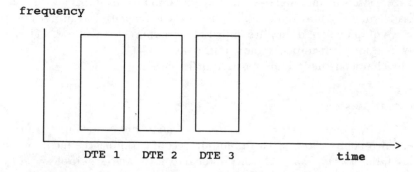

Figure 6.17 *Baseband*

frequency

| channel 1 |
| channel 2 |
| channel 3 |

time

Figure 6.18 *Broadband*

In the case of fibre optics, FDM is called wave division multiplexing (WDM). With broadband transmission, it is therefore possible to have multiple, independent, and concurrent transmission channels.

6.6 MEDIUM ACCESS CONTROLS

With baseband transmission all the DTE must share a single transmission path, unlike broadband transmission for which multiple channels are available (Figure 6.19). Consequently there is the need for discipline with respect to access and use of the transmission path that must be imposed on all network users. The two standard techniques that have been adopted are carrier sense multiple access with collision detection (CSMA/CD), which is used with bus topologies and is also known as Ethernet, and control token, which is used with either bus or ring topologies.

6.6.1 Ethernet

Ethernet, the most widely used local area network, is a multidrop, bus or tree configuration that normally uses baseband TDM. As such it is a cheap and simple method for communication between locally distributed computers. This is a shared communications channel in which data packets, each with a destination

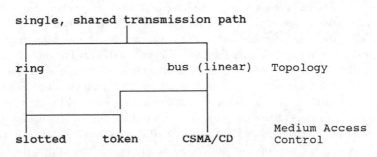

Figure 6.19 *Medium access control*

```
data rate : 10Mbps
maximum segment length : 500m
maximum end to end length : 2.5km
data encoding scheme : Manchester
```

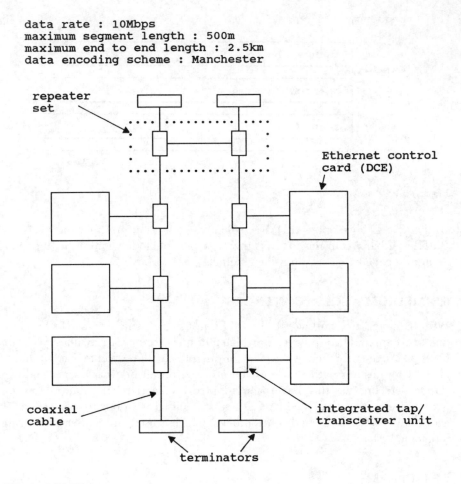

Figure 6.20 *Ethernet*

address, are broadcast (all frames are seen by all computers). There is no central control point and channel access is by a statistical arbitration scheme. The network can have a maximum segment length of 500 m; however, the use of up to 4 repeaters gives a maximum total length of 2500 m (Figure 6.20). The data rate depends on the type of cable used. The Ethernet transmits data in groups or blocks called either *packets*, *data frames* or *information frames* (Figure 6.21). A data frame will contain not only user data but additional information for control purposes. The frame must be at least 64 bytes long. At transmission, if the data in the data field are less than 64 bytes, the Ethernet controller will pad the field with zeros up to the correct 64 bytes. The frame includes both the destination and the source address. Every piece of DCE receives all frames, but the data are read only if the frame has the appropriate address. The source address allows the recipient to reply. Error detection is by CRC.

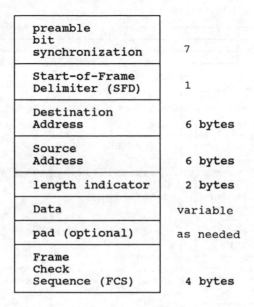

preamble bit synchronization	7
Start-of-Frame Delimiter (SFD)	1
Destination Address	6 bytes
Source Address	6 bytes
length indicator	2 bytes
Data	variable
pad (optional)	as needed
Frame Check Sequence (FCS)	4 bytes

Figure 6.21 *Ethernet frame*

6.6.1.1 Packet Transmission and Reception. In this communication system, there is only one cable that is used by many computers. There must therefore be a control mechanism to ensure an equitable use of the medium. The control mechanism is an access method called carrier sense multiple access/collision detection (CSMA/CD), also known as Ethernet, which is implemented by Ethernet control cards. A computer that wishes to use the network must contend for use of the cable (the Ether) until it acquires it; the computer can then send data. To contend for sole use of the cable, the computer listens to see if the cable is busy (carrier sense) and will defer transmission until the cable is silent. The deferring station will then start to transmit. During transmission, the computer must listen for a collision as other stations may also have started to transmit during this quiet period (multiple access).

The Ethernet specification supplies the functions of the lowest two layers of the OSI model, but as such may have multiple protocols above it to enhance the functionality of the network. A variety of other higher layers can exist. It is the responsibility of the upper layers to address issues such as acknowledgement and retransmissions. Bit synchronization is achieved by the first 64 bits of the data frame—the preamble.

The conditions for a controller to accept a frame are that: the destination address matches the station address (*i.e.*, each station has a unique address); the destination address is a broadcast address (*i.e.*, the data frame is addressed to all stations); the destination is a multicast address (*i.e.*, a group of machines share a common address); and the station is set in 'promiscuous' mode in which the controller will accept all frames irrespective of the destination address.

Computer bus

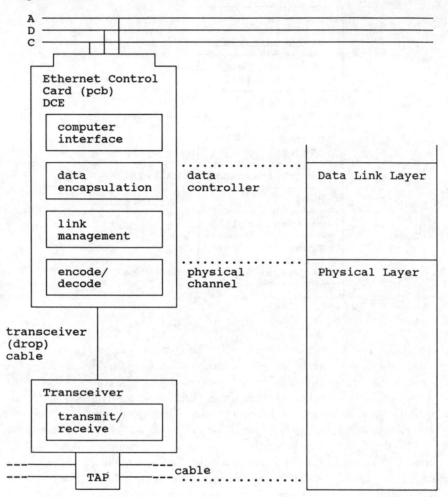

Figure 6.22 *Ethernet controller*

Should the address not be suitable, the rest of the frame is ignored. If the address is correct, the receiver accepts the frame into a buffer where it is checked for errors by the CRC. If an error has occurred the frame is discarded. It is important to realize that the receiving station will not notify the transmitting station that the data frame has not been correctly received. This is called *connectionless*-oriented protocol. Indeed, the receiving station may even be switched off. Further, the controller may be busy and have missed a data frame. Hence the need for other functionality that must be provided by higher layers. The communications subunit consists of an Ethernet controller (Figure 6.22 and 6.23) and an associated transceiver (transmitter/receiver) unit and provides the following displayed functions.

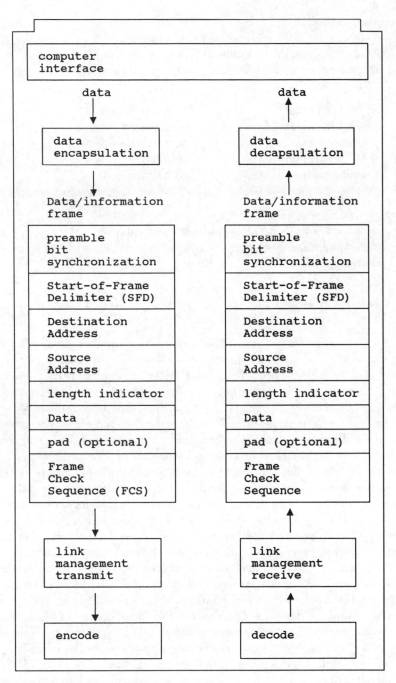

Figure 6.23 *Ethernet controller functional units*

Ethernet controller:
computer interface (serial/parallel conversions)
data encapsulation/decapsulation (packet generation)
link management (CSMA/CD algorithm implementation)
encoding/decoding

data encapsulation	data decapsulation
preamble generation	synchronization
address generation	address recognition
CRC generation	CRC check
padding as appropriate	padding removal

Transceiver:
cable connection (TAP)
transfer data frames between controller and cable
carrier sense (CS)
collision detection (CD)
jabber control

The Ethernet controller provides the functions associated with the lower two layers of the OSI model, *i.e.*, the physical and data-link layers.

6.6.1.2 Collision Detection. Stations do not indicate that they are about to transmit. With this mode of operation it is possible for two items of DCE to attempt to transmit data at the same time. To prevent this occurring, each DCE item listens to the cable to detect whether or not a frame is currently being transmitted. If a carrier signal is sensed (CS), the DCE delays transmission until the cable is free. It is, however, still possible that two DCE items may sense a free cable and transmit simultaneously. Collisions therefore occur immediately after the start of transmission during what is called the *collision window* or *collision interval*. After this initial period, one station has control or has acquired the cable (Figure 6.24). No acknowledgement is sent from the recipient Ethernet card on successful receipt (or not) of the data frame. This must be done by higher-level protocols. The collision window is a function of the end to end propagation delay. It takes a minimum of a 64 byte frame to be transmitted on a 2500 m cable (including repeaters) to ensure that all stations have seen the packet.

The transceiver can detect collisions (collision detection). To allow for this contingency, each sender simultaneously monitors the cable when transmitting a data frame. If the transmitted and monitored signals are different, a collision has occurred. If a station detects a collision, transmission of the rest of the data frame is aborted (Figure 6.25). Furthermore, to ensure that all stations are aware of this collision, a jam sequence is transmitted for a short period of time. The jam sequence is a random series of data 4 to 6 bytes long. Each station then invokes a back-off algorithm that involves a delay of a short random period before retransmission is attempted. Should another collision occur, there is a progressive increase in the time delay between attempts. There is a maximum number of retransmission attempts.

The analogy here would be a dinner party. You can only speak to someone if

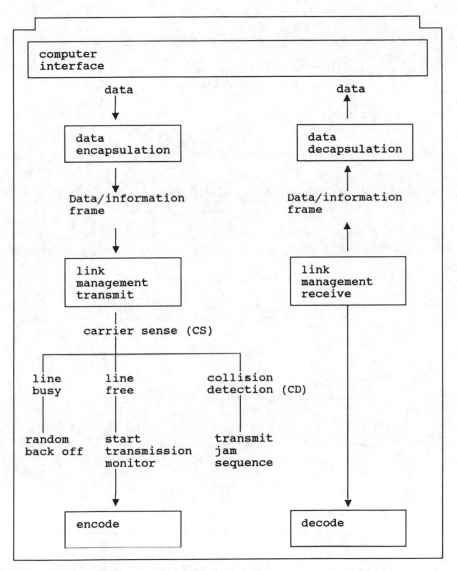

Figure 6.24 *Ethernet controller functions*

there is a lull in the conversation. It can be concluded that this type of medium-access control is probabilistic—transmission within a given time period cannot be guaranteed. This can be an important consideration depending on the application requirements.

6.6.1.3 Ethernet and the IEEE CSMA/CD Specification. Ethernet was first implemented in 1975 with the 'Ethernet specification'. Ethernet V1.0 was published by a group of companies (Xerox, Intel, and DEC) in 1980, and the V2.0 in 1982. The IEEE defined the IEEE 802.3 standard 1983, which was accepted by ANSI in 1984 and the ISO in 1985.

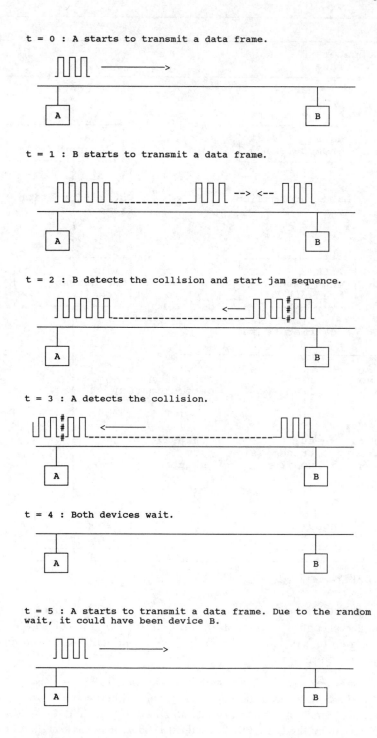

Figure 6.25 *CSMA/CD sequence*

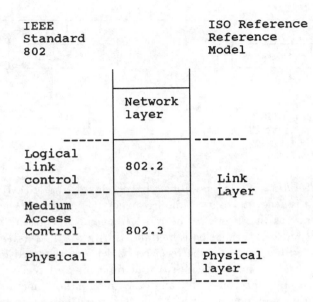

IEEE 802.3 : CSMA/CD.

Figure 6.26 *IEEE/ISO Standards*

There are some differences between the Ethernet and IEEE 802.3 standards (Figure 6.26). The IEEE 802.3 standard has two sublayers, the media access control (MAC) layer and the logical link control (LLC) layer. Ethernet combines both of these sublayers into a single protocol. Only an unacknowledged connectionless protocol is provided by Ethernet; all packets are transmitted on a best effort basis. The IEEE 802.2 specification for the data link layer provides two classes for service, type 1, a connectionless protocol, and type 2, a connection-oriented protocol. The IEEE 802.3 standard can make use of the 802.2 data-link service to provide not only the connectionless service, but also a connection-oriented service; all packets are guaranteed to be delivered.

Two types of service are provided by the LLC. The unacknowledged connectionless service provides the user with a data-transfer service with the minimum of protocol overheads. With this type of service functions such as sequencing and error recovery are provided in a higher protocol. With the connection-oriented service a link level logical connection is established before any data transfer occurs and error recovery and sequencing is implemented.

In the type 1 connectionless service, data are transmitted onto the cable regardless of whether or not the receiving station receives the packet. The best analogy is sending a letter. This protocol depends upon the networking software in the upper layers and the functions of the data link layer.

In the type 2 connection-oriented service, a logical connection is established between communicating computers before data are sent. Sequence numbers are applied to the packets to ensure correct ordering.

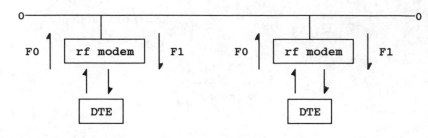

Figure 6.27 *Rf modems*

6.6.1.4 Broadband Ethernet. It is possible to have a broadband (FDM) LAN. With broadband, it is necessary to use radio frequency (rf) modems. They are called radio frequency as they work in the radio frequency spectrum. The selected carrier frequency for transmission is modulated with the data to be transmitted and the selected frequency for the received data is demodulated to obtain the data (Figure 6.27). The bandwidth of the transmit and receive frequencies is determined by the bit rate and the modulation method used. A typical high data rate modem, such as one that would be needed by a 10 Mbps Ethernet channel, transmits from 54 to 72 MHz (F0) and receives from 210.25 to 228.25 MHz (F1) when a single cable is used (Figure 6.28). A dual-cable system uses one cable for transmission and one for reception. In this case, the same frequency is used for transmission and reception, typically 54 to 72 MHz (Figure 6.29).

There remains a lot of bandwidth. Standard frequency, 6 MHz channels have been specified for LAN systems (and community antenna television, CATV) (Figure 6.30).

There is, however, a price to pay for having this large number of data channels based on one cable and that is the relatively high cost of the rf modems. It should nevertheless be noted that an additional advantage of broadband coaxial transmission is that it can be used over longer distances than baseband. Broadband

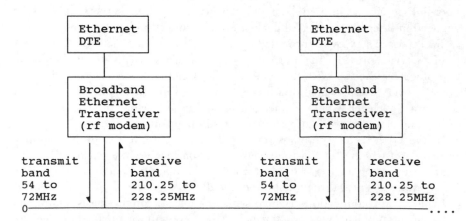

Figure 6.28 *Single cable broadband system with a 10 Mbps Ethernet*

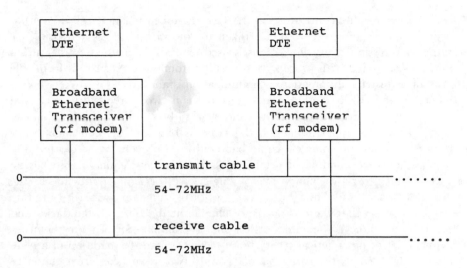

Figure 6.29 *Dual-cable broadband system with 10 Mbps Ethernet*

cabling therefore tends to be used as a primary transmission medium in the manufacturing industry when the separation of buildings is quite large. In addition, the other channels are used for services such as closed-circuit television and voice communications.

6.6.2 Token-controlled Networks

The second MAC method uses the control (permission) token that is passed between DTE according to a defined set of rules. A DTE item may only transmit a frame when it possesses a token. After data-frame transmission the token is passed to another DTE item. There are two main implemented standards of this method, token ring and token bus. We will consider only token bus, but will omit the details concerning issues such as token loss and regeneration.

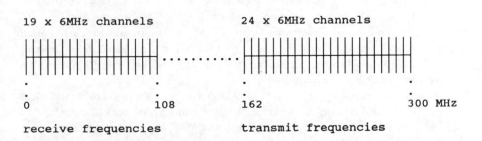

Figure 6.30 *IEEE 40.1 frequency standard for 6 MHz channels on a broadband coaxial system*

6.6.2.1 Token Bus. By contrast to CSMA/CD, the token bus network is deterministic in nature due to the use of a token to control bus access and prioritize
frame transmission. This type of bus is used extensively in the manufacturing
industry concerned with process control and automation. As such, it forms the
basis of industrial automation communication standards. The token bus is
defined by the IEEE standard 802.4. Token bus networks typically use coaxial
cabling and operate in broadband and a slightly modified baseband mode
called carrierband. Carrierband is the same as baseband in that each transmission occupies the complete cable bandwidth. However, in the carrierband
mode, all data are modulated before transmission using frequency shift keying
(FSK), which encodes binary 1 as a single cycle of signal frequency at the
normal bit rate, and binary 0 as two cycles of signal at twice the bit rate
frequency. Note that there is no phase change at the bit-cell boundaries and
that this technique is sometimes called phase coherent FSK. Factory environments are electromagnetically very noisy. The baseband is made up of a wide
bandwidth, whereas the carrierband has only two frequencies. It is therefore
possible to use filters to permit only these two frequencies to be received. This
results in a significant improvement in the signal-to-noise ratio. This obviously
cannot be done with baseband.

The broadband mode has several channels (up to 10 Mbps each) while
single-channel media use carrierband modulation. We have therefore a broadcast coaxial with several channels (each 10 Mbps) for the main channel (spine),
and a single channel carrierband (modified baseband) for subnets.

There is a standard physical interface module (PIM) that contains modulation
and interface control circuits responsible for modulation and encoding, de-
modulation and decoding, and clock generation.

6.6.7.2 Token-bus Operation. There is a single control token, possession of which
allows frame transmission. The DTE is linked to form a logical ring, even though
broadcasting is used. The token is passed from the logical predecessor (upstream)
to the logical successor (downstream) (Figure 6.31). Each DTE item only needs
to know the address of the successor, *i.e.*, there is explicit ordering for token
passing.

6.6.2.3 Priority Operation. It is possible to implement a priority mechanism using
the token bus. The access method recognizes four access classes suitable for
manufacturing automation and process control: class 6 for urgent, top priority
messages, critical conditions, and alarms; class 4 for messages associated with
normal control actions, and ring management functions; class 2 for messages
associated with data logging; and class 0 for messages associated with general file
transfer, and program downloading.

The medium capacity (bandwidth) is shared between all DTE and it is
essential that transmission of low-priority frames must not block the transfer of
high-priority frames. Similarly, the transmission of high-priority frames must not
result in starvation of low-priority frames.

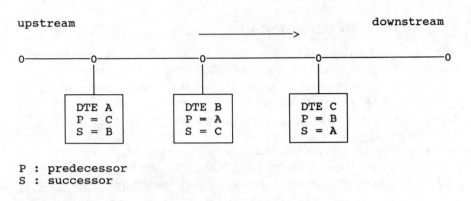

P : predecessor
S : successor

Figure 6.31 *Token bus*

6.7 LOCAL AREA NETWORK AND THE OSI

The protocol standards for LANs in the context of the OSI model are defined in IEEE Standard 802 (Figure 6.32). The three MAC standards and the associated physical media specifications appear in the following IEEE standards documents: IEEE 802.3, CSMA/CD; IEEE 802.4, token bus; and IEEE 802.5, token ring.

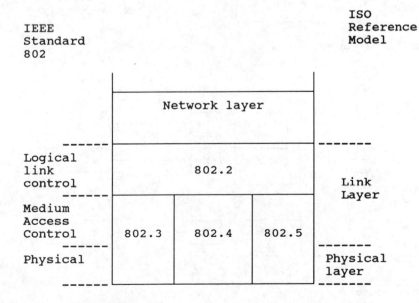

Figure 6.32 *IEEE/ISA Standards*

6.8 TERMINAL-BASED NETWORKS

Terminal-based networks are characterized by normally having a single central computer to which access is by terminals (Figure 6.33). The terminals can have different degrees of intelligence or local processing ability. Simple interactive terminals will have a simple UART with asynchronous character mode protocol, with error control by echo checking and flow control by X-ON/X-OFF. More complex terminals would operate in synchronous block mode with error control. Remotely situated terminals can be connected by the use of modems (Figure 6.34). In order to reduce cabling costs, it is possible to provide a single, high bit rate line between remote locations that is shared by the lower bit rate terminals (Figure 6.35). The device that achieves this is called a terminal multiplexer.

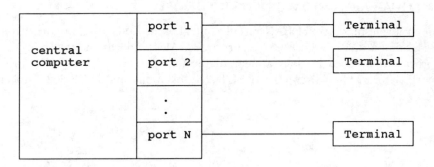

Figure 6.33 *Terminal based network*

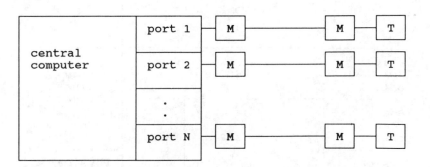

M = Modem
T = Terminal

Figure 6.34 *Terminal based network with modems*

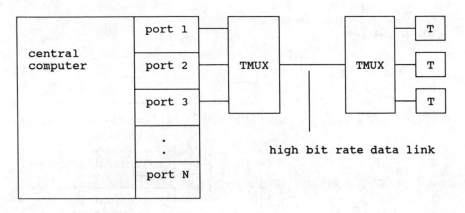

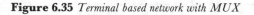

TMUX = Terminal Multiplexor

Figure 6.35 *Terminal based network with MUX*

6.9 OSI ENVIRONMENTS

Within a discrete OSI environment (OSIE), the specific application services suitable to that environment must be selected along with the different modes of operations and classes of service provided by the lower network-dependent layers (Figure 6.36).

A number of OSIEs based on the OSI model exist. The Manufacturing Automation Protocol (MAP) is a set of protocols selected to achieve OSI within an automated manufacturing plant (Figure 6.37). Manufacturing Automation Protocol is a factory-wide backbone cable distribution network. This coaxial cable operates at 10 Mbps using the broadband mode to support the wide range of communication requirements. The MAC sublayer is 802.4, the token bus, selected because of its deterministic nature. The network layer and the LLC sublayer operate in connectionless mode with the transport layer operating in connection-oriented mode. The distributed information services include FTAM, MMS, and DS. Communication within a manufacturing cell (the laboratory) allowing data exchanges between work stations uses the simpler and cheaper carrierband transmission. Similarly, the Technical and Office Protocol (TOP) is a selection from the OSI model to achieve OSI within a technical and office environment. In this case the transmission medium is also a 10 Mbps coaxial cable in baseband using the 802.3 (CSMA/CD) MAC protocol. As with MAP, the network layer and LLC sublayer operate in connectionless mode, while the transport layer operates using class 4 service. The distributed information services include FTAM, MHA, JTM, and VT.

6.10 CLIENT SERVER ENVIRONMENTS

One of the main driving forces behind the proliferation of LANs is the potential to share expensive resources with many users running different applications. This

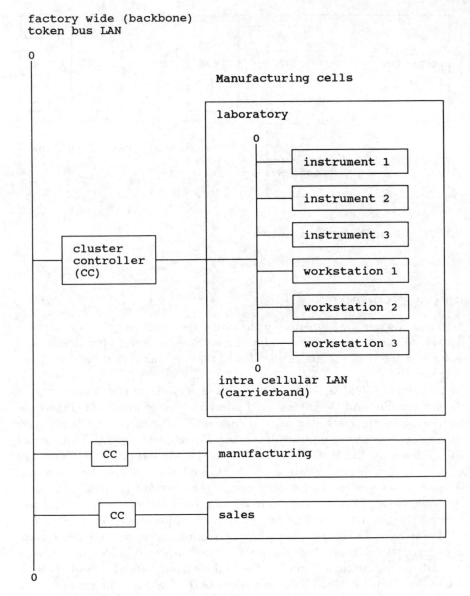

Figure 6.36 *An OSI environment*

is achieved by means of a *LAN server*. A LAN server provides shared access to a resource (Figure 6.38). On the network there can be numerous PCs, not necessarily with the same operating system, that can access the LAN server. Typically, a LAN server is a dedicated computer that may control one or more resources. The user can request an application program from the server, which can then be run from the local PC. The file server may accept multiple connections from stations. The server may or may not be transparent to the user. We note that the

Application Process

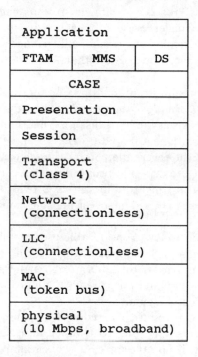

Figure 6.37 *Full MAP*

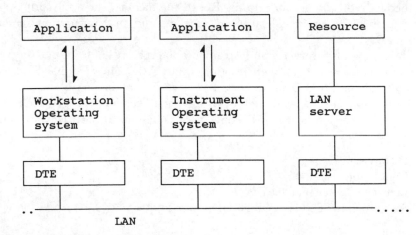

Figure 6.38 *Client server*

modularity of this approach ensures a simple and cheap increment to the resources that are available on a given network; more users and resources can easily be added.

Servers make it possible for there to be common files for all users; a *file server* administers these shared files. A file server is a specific type of LAN server that provides a shared storage space suitable for file sharing. As we will see in Chapter 7, Introduction to Data Processing, data must be managed. Typically, this is done by the network manager. Services provided by the file server include a back-up service in which files are periodically copied onto another disk or magnetic tape, and recovery procedures that allow restoration of the system should the file server fail.

The term file server has no strict definition. It can be taken simply to mean *shared disk capacity* or *full shared file management.* When the file server is used for shared disk capacity, it acts as the hard disk or even as an extension to the disk capacity on each workstation. Automatic back-up and the shared larger disk provides a cost/benefit; however, there is no file sharing, and files must be transferred between users by specific programs. This can be constrasted with shared-file management in which the file server provides, in addition to extra disk space, multiple access to files. The server manages concurrent file access and access rights.

Certainly this approach allows considerable user mobility. This provides an undoubted advantage in some laboratories where the users may work on different computers in different parts of the laboratory. Furthermore, there is a degree of 'graceful degradation'. Should one, or more application computers fail the network will still be able to function.

Use of the network can result in a reduced response time for the system; there may be many users all trying to transfer files across the network. To alleviate this problem it is possible for each workstation to have a *file cache.* A file cache is a memory location of limited capacity that holds recently used files thereby reducing network traffic. Recall the memory hierarchy principle. Other types of server include printer, modem, and instrumentation servers (Figure 6.39).

A chromatography server is an instrumentation server. For example, the VG chromatography server is a high resolution, fast A/D converter that connects directly to the network. Depending on the server it may be possible to acquire data from more than one instrument by means of independent ports. As such, this type of server has 'on board' A/D converters, communications hardware, and memory.

A/D converter
    Conversion resolution: 22 bits
    Linearity: better than 0.01%
    Sampling rate: 800 Hz with 50 Hz power supply

Network
    IEEE 802.3 (CSMA/CD)

Memory
    1 Mb RAM

The use of chromatography servers is an alternative approach to the use of chromatography boards that can be plugged directly into PCs, which in turn run specialist software accordingly.

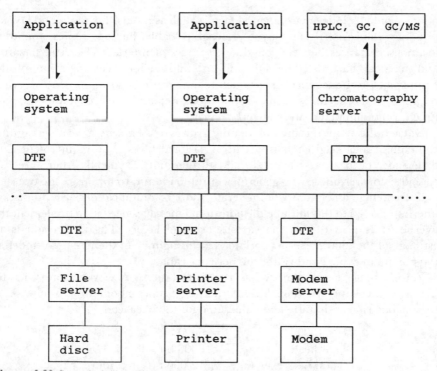

Figure 6.39 *Server including printer, instrumentation, and modem servers*

6.10.1 Structured Query Language Front Ends

Single user, single PC databases are now commonplace. The use of a shared database requires PCs to be networked. The database then resides on the file server allowing multiple-user access. In order to make extensive use of such a system, the entire database has to be moved over the network to your PC. If all users tried to do this at the same time, the network response time would suffer. Structured query languages (SQLs) address the problems of accessing data from various sources. The SQL software resides on the server and responds to SQL commands transmitted over the network; this results in a significant reduction in network traffic.

Though simple to use, SQL can become difficult when queries are complex. To address this problem, some PC application packages have SQL functionality. The user may then work in the chosen application with SQL commands being generated transparently. Alternatively, there are graphically-based tools specifically available, *e.g.*, Gupta's Quest.

6.11 LOCAL AREA NETWORK SELECTION AND INSTALLATION— A GUIDE

The geography of the site must be studied, the site audit. For a multi-building/ room installation a standard questionnaire/check list can be of value. Considerations should include environmental factors such as room dimensions and layout, building dimensions and layout, possible cable/ducting routes, building locations, distances between buildings, sources of interface/electrical hazards, and physical hazards. The site audit should include consideration of the equipment to be connected (model, configuration, location *etc.*), services required, response times (some applications demand an acceptable response time), expansion plans, and network security.

Data-traffic analysis is used to identify groups of users who have a demanding communications need compared with occasional users. The requirement is to determine the peak and average data throughputs. Typically, there are two categories of user, those who generate continuous data streams (*e.g.*, file transfer, and archiving) and those who generate short communications (*e.g.*, database queries). Added to this must be the additional messages needed to implement the overheads associated with the various protocol layers. The final result is an estimate of the total mean network load. This figure can then be used to define the minimum cable bandwidth, the network capacity.

For example, the laboratory has 100 workstations. A workstation generates messages involving six 1000 bit information frames about ever 5 min. As a reasonable 'rule of thumb', allow the same for the overheads.

$$
\begin{aligned}
6 \times 1000 &= 6000 \text{ bits} \\
\text{Overheads} &= 6000 \text{ bits} \\
\text{Total data exchange} &= 12\ 000 \text{ bits in 5 min} \\
&= 40 \text{ bps}
\end{aligned}
$$

Even with, say, 10 concurrent users, we are well within the bandwidth capacity of our link. It is important that we are well below the upper limit, otherwise the network service, and particularly Ethernet, will rapidly deteriorate in terms of response time and will be subject to long delays.

After selection and implementation of the network there must be acceptance testing. Factors that should be considered include flood testing (excess traffic), stress testing (vibration, electrical, environmental), and facility testing.

The operation of the network on a routine basis is called network management. Tasks include administration, configuration, billing, fault diagnosis, traffic analysis, maintenance, and back-up strategy.

6.12 SELECTION AND INSTALLATION OF EQUIPMENT AND DATA COMMUNICATIONS CABLES

It is essential to adopt a systematic approach to the selection and installation of cables in order to minimize environmental noise and possible cable damage; this is capable planning. Connections between devices are one of the major problems in computer systems. The following points should be noted.

Ensure that the type, size, and length of cable are correct for the specification requirement. For example, the RS 232 specification defines suitable data transfer rates, cable runs (*i.e.*, length of cable), cable terminations, *etc.*

Analyse the working environment for possible sources of noise. One of the main causes of EMI is electromagnetic induction whereby power cables running near to or parallel with communication lines induce noise.

Analyse the proposed cable route for sources of noise and eliminate the source or re-route the cable.

Consideration must be given to the entire cable plan. This is the cable-layout scheme for power, lighting, communications, *etc.* As far as possible these cables should be segregated and isolated. Induced voltage is a function of the square of the distance from the source. In particular, data cable should be contained in metal ducts. When power cables must cross data cables, try to ensure that they do so at an angle of 90°.

Analyse the proposed cable route for potential mechanical or chemical hazards. Note that cable routes are often difficult to access and the cabling is therefore often 'fit and forget'. It is important that the cable is not exposed to a deleterious environment.

Where possible, the cable route should be accessible allowing for easy inspection and maintenance.

Use published codes of practice and standards for earthing. In particular, the screen should be earthed at one end and the screen continuity should be maintained at the termination.

If it is not possible to remove all sources of noise, implement suitable protective measures such as shielding.

As mentioned in Chapter 2, the use of a UPS is an aid to a 'clean' power supply.

6.13 DEVELOPMENTS

Very often computer networks must be installed and used in buildings with no cabling infrastructure. The rooms cannot be modified, for whatever reason, to allow the use of ducting and specified ducting routes. Wireless networks are a possible solution. Furthermore, computers on a wireless network may be simply unplugged from the mains and moved to a new location. There are three main transmission media, radio frequency, microwave, and infra-red. Radio frequency based networks send signals to other machines by small transmitters, either to a central hub for retransmission or peer to peer. Depending on the strength of the signal, distances of up to 250 m are possible. The use of transmitters allow coverage of a complete site. Two principal techniques are used, wide spectrum, low amplitude, and narrow spectrum, high amplitude. The wide spectrum (or spread spectrum) technique sends signals across a number of different frequencies. The main problem with the wide-spectrum technique is that the frequency spectrum must be shared by other users; bandwidth is therefore allocated. The 902 MHz to 928 MHz band used in the US has been assigned to cellular telephones operating at 928 MHz within the UK. The European Telecommunications Standards Institute has developed the Digital European Cordless Telecommunications Standard (DECTS) for the narrow-spectrum technique, which operates from 1880 MHz to 1900 MHz. Standards for wireless LANs are currently under consideration by the IEEE 802.11 committee.

The use of low-powered microwaves has been largely pioneered by Motorola. Microwave LANs operate at about 18 GHz providing fast data-transmission rates. Furthermore, the use of intelligent transmitters, which can determine the best transmission route, helps ensure good communication links.

Infrared LANs must use 'line of sight', in which a single beam must pass from the transmitter to the receiver, rather like the remote control for a television. This is a simple solution, but the price to pay is the difficulty of ensuring an uninterruptable line of sight—people may pass in front of the beam and corners cannot simply be negotiated except with mirrors.

CHAPTER 7

Introduction to Data Processing

Let us now try to relate the complex data that are typically used in the 'real world laboratory' to how they are stored and manipulated by a computer. At the machine level all we have is a sequence of bytes in the memory. As we have seen, these bytes can represent different types of data such as ASCII characters, integers or floating-point numbers. By using variables we have a symbolic representation of the named universe of discourse.

7.1 INTRODUCTION—RECORDS AND FILES

In everyday life data are organized into collections of related facts often recorded on index cards or *record* sheets (Figure 7.1). A record is a useful collection of related data items that are treated as a single unit for purposes of storage, access, and processing. Records have *data fields*. A data field, or simply a field, contains a single data value such as time, pH, *etc*. Each data field is in effect a variable and

```
Date : integer

Work_number : integer

Customer_number : integer

Customer_name : ASCII characters

Customer_address : ASCII characters

Sample_number : integer

Sample_material : ASCII characters

Test_Number : integer

Test_Name : ASCII characters

Test_result : floating point

Analyst : ASCII characters
```

Figure 7.1 *A samples worksheet record*

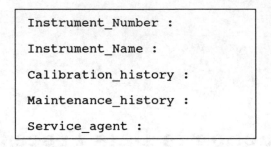

```
Instrument_Number :

Instrument_Name :

Calibration_history :

Maintenance_history :

Service_agent :
```

Figure 7.2 *An instruments record*

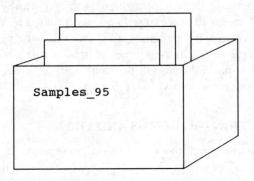

Figure 7.3 *File of records called Samples__95*

must have a data type. In programming, a record is a data structure that allows a collection of variables to be treated as a single unit even though the variables are of different data types. Consider for example a laboratory worksheet. There could be many variables on such a worksheet, but for simplicity and brevity only a few are included.

We can have different types of record, which for identification purposes have different names. For example, we may have a collection of records for instrument maintenance (Figure 7.2). A collection of similar records is called a *file*. Files are designed for handling large volumes of data. Index cards will be stored in an index card holder; larger records may be stored in the drawer of a filing cabinet

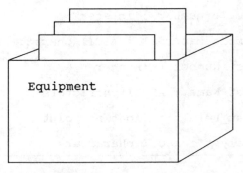

Figure 7.4 *File of records called Equipment*

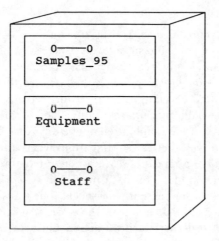

Figure 7.5 *A collection of files*

(Figure 7.3). A file may then be treated as a single unit for ease of manipulation. Files can be identified by name. The file Samples—95 holds records all of the same type, in this case called Samples—worksheet. Similarly, we can have a file of instrument records; in this case we will call it the Equipment file (Figure 7.4). We now have a file collection that mimics our traditional filing cabinet (Figure 7.5). However, in our case the files are stored on magnetic disk.

Irrespective of the type of record, the data in the fields can be classified as control, static, or dynamic data. For control data, each record must have a unique identifier, sometimes called a *primary key* (*e.g.*, work—number). Using this key it is possible to access only one record. Static data, as the name implies, are relatively static. In our example, this would be all the information that relates to the sample but does not include the test results. Modification of this type of data field is called *amendment*. Dynamic data are the generated data, *i.e.*, test results. Modification of this type of data field is called an *update*.

There are many different types of file such as reference, report, dump, historical, *etc.* Two common types are master files, which hold relatively permanent records, and transaction files, which are made from source documents and are used to modify master files.

Master file: permanent records
Transaction file: made from source documents/data; used to
                    modify/update master file

7.2 FILE STRUCTURES

Computer files may be structured in different ways. The file structure determines the performance of file access and manipulation. The three main types of file structure that we will consider are sequential files, index sequential files, and direct access or hash files.

7.2.1 Sequential Files

A sequential file is a collection of records that are stored in *sequential order*. This logical organization mimics the file-storage characteristics of magnetic tape. Data processing is sequential only; records can be examined one at a time and only in sequential order. Consequently, if it is the last record that is required it can only be accessed after checking all preceding records. For interactive demand processing this would give unacceptably poor performance. Imagine trying to look up Zybinsky in the London telephone directory by going through all preceding names! However, there are applications where this type of processing is appropriate. In these applications, data generated throughout the day are collected in a transaction file. This data may be new worksheets that have been generated or test results for current worksheets. The transaction file is then used to update the master file that holds all the worksheet records. Master-file updating consists of first ordering the transaction file according to the same criteria as the master file (ordered by primary key). Each record from the transaction file is read and the master file is examined for a match. In effect, both files run in unison. The advantage is that the sequential access time between adjacent records is extremely fast. For large data volumes this is an efficient method of data processing. The update of the master file is usually done overnight or as a background job. Sequential-access files may be stored and processed using either magnetic tape or hard disks.

7.2.2 Index Sequential Files

Index sequential files address the slowness of the sequential file. The sequential nature of the records is maintained, but an index is also used. The index segments the records and greatly reduces the time taken to access an individual record. The London telephone directory is divided into separate volumes, *e.g.*, A to B, C to D, *etc.* To find Zybinsky we go directly to the volume Y to Z and we then search sequentially. Very large data volumes require multiple levels of index. Index sequential files give a good performance for handling large data volumes owing to the sequential organization, but random searches are also fast because of the index. The disadvantage is that the index itself requires storage space. Index sequential files can only be stored on hard disks.

7.2.3 Direct-access or Hash Files

Direct-access or hash files use the full performance of hard disks. The primary key is converted or hashed to a number that refers directly to a disk address (cylinder number, surface number, and sector number). It is possible to go directly to the required record with these files and it is these that give the best performance for random files enquiries. However, the hashing procedure is such that the records are placed all over the disk; the next record in sequence would not be next to the current one. The updating of large files would be inefficient due to the overhead of the disk seek time (Table 7.1).

Table 7.1 *Device performance*

Device	Average direct-access time	Sequential-access time/ms
Magnetic tape	1–3 min	2–10
Magnetic disk	10–50 ms	5–10

7.3 FILE-PROCESSING CRITERIA

The efficiency of file processing depends on the nature and requirements of the given application. Considerations include the following: file size, or the total number of records; activity, or the number of records that need processing in a given processing period; response time, or the elapsed period between the demand for processing and fulfilment of that demand (Is the application demand or batch driven?); volatility, or the proportion of records added to or deleted from a file over a given processing period; integration, or the extent to which files are interdependent; and security, or back-ups and recovery.

A *hit* is a record that needs to be processed/accessed and

$$\text{hit ratio (hit rate)} = \frac{\text{number of records accessed in a run}}{\text{total number of records in the file}}$$

$$\text{fan in/out ratio} = \frac{\text{number of accesses to the file}}{\text{number of records hit}}$$

$$\text{volatility} = \frac{\text{number of additions} + \text{number of deletions}}{\text{number of records in the file at the start}}$$

7.4 FILE MANIPULATION AND MANAGEMENT

A file-management system is the system software that provides all the necessary file manipulation services to users. Files may be manipulated as a unit using the following operations.

Open, *i.e.*, data in the file may be referenced. This is analogous to opening a filing cabinet.

Close, *i.e.*, data in the file may not be referenced. This is analogous to closing the filing cabinet.

Create, *i.e.*, create a new file. This is analogous to getting access to another filing cabinet drawer.

Delete, *i.e.*, delete a file. This is analogous to emptying the contents of the filing cabinet drawer into the trash can.

Copy, *i.e.*, copy the file contents onto another file name. This is analogous to transferring the contents of one drawer to another.

Rename, *i.e.*, change the name of the file. This is analogous to changing the name on the drawer.

List, *i.e.*, display the file contents. Remove the contents of the drawer for access.

The individual data items (records) within the file may be manipulated using the following operations.

Read, *i.e.*, data are read from the file. This is analogous to reading a record.

Write, *i.e.*, data are written to the file. This is analogous to writing data onto a record.

Insert, *i.e.*, add a new data item from the file. This is analogous to putting a new record in the file.

Delete, *i.e.*, remove a data item from the file. This is analogous to removing a record from the file.

7.5 SEQUENTIAL FILE PROCESSING IN PASCAL

As we saw briefly in Chapter 2, using Pascal it is possible to read data from the keyboard, process the data, and write the data to the screen.

```
PROGRAM Moles (Input, Output):
VAR moles, litres, molarity : INTEGER;
BEGIN
     WRITE('Please enter moles and litres');
     READ(moles);
     READ(litres);
     moles = molarity * litres;
     WRITE(moles);
END.
```

In this program, moles, litres, and molarity are variables of *data type* INTEGER, the contents of which can be changed. However, the contents must be of type INTEGER, a simple binary number. INTEGER, FLOATING__POINT, and CHARACTER are data types known to the computer. If they were not known to the computer the type would have to be defined. In effect, the definition of the data type has been done for you.

```
PROGRAM Moles (INPUT, OUTPUT);
(TYPE INTEGER = simple binary number;)
VAR moles, litres, molarity : INTEGER;
BEGIN
     WRITE('Please enter moles and litres');
     READ(moles);
     READ(litres);
     moles = molarity * litres;
     WRITE(moles);
END.
```

We wish to use a variable called Samples__worksheet that is a record. But what data type is this? We must define it ourselves. The TYPE declaration is used to define a new data type. We have a record of type Samples__record with a user-defined collection of data fields.

```
PROGRAM File_processing (INPUT, OUTPUT);
TYPE  Samples_record = RECORD
                        Date                : integer;
                        Work_number         : integer;
                        Customer_number     : integer;
                        Customer_name       : ASCII characters;
                        Customer_address    : ASCII characters;
                        Sample_number       : integer;
                        Sample_material     : ASCII characters;
                        Test_number         : integer;
                        Test_name           : ASCII characters;
                        Test_result         : floating point;
                    END;

VAR   Samples_worksheet : Samples_record;
```

The variable Samples_worksheet is of data type Samples_record. The choice of the type name Samples_record is made by the programmer.

In standard Pascal, input (from the keyboard) and output (to the screen) is performed through external files identified in the program heading (INPUT and OUTPUT). Additional or alternative files may also be listed in the program heading. Recall that one of the files we are going to use is called Samples_95. All files listed in the heading, with the exception of INPUT and OUTPUT, must be declared as file variables.

```
PROGRAM File_processing (INPUT, OUTPUT, Samples_95);
TYPE  Samples_record = RECORD
                        Date                : integer;
                        Work_number         : integer;
                        Customer_number     : integer;
                        Customer_name       : ASCII characters;
                        Customer_address    : ASCII characters;
                        Sample_number       : integer;
                        Sample_material     : ASCII characters;
                        Test_number         : integer;
                        Test_name           : ASCII characters;
                        Test_result         : floating point;
                    END;
      Samples_file = FILE of Samples_record;

VAR   Samples_worksheet : Samples_record;
      Samples_95 : Samples_file;
```

We now have a variable called Samples_worksheet that is of the data type we defined, *i.e.*, Samples_record. We have a file called Samples_95 that we have defined to be of data type Samples_file, which can hold only records of data type Samples_record. We will now look at sequential file processing using the language Pascal.

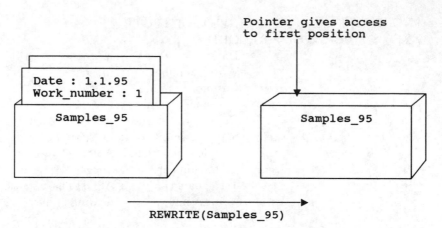

Figure 7.6 *REWRITE*

7.5.1 Opening a File

Prior to use, a file must be opened, REWRITE(filename) opens a file for writing, and the file pointer is then positioned at the beginning of the file. In our example, this would be REWRITE(Samples__95). We can now write records to the file. Using our index card analogy, we have just emptied out any index cards and the box is empty (Figure 7.6). RESET(filename) opens a file for reading, and the file pointer is again positioned at the beginning of the file, but this time we have access to the first record (Figure 7.7).

7.5.2 Writing Records to a File

Records can be written to a file by using WRITE(Filename, Recordname), which places the record into the file and increments the file pointer. If the file is

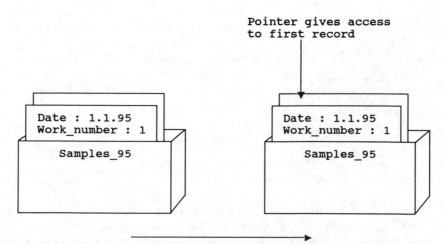

Figure 7.7 *RESET*

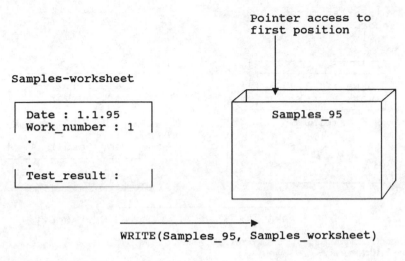

Figure 7.8 *WRITE(Filename, Recordname)*

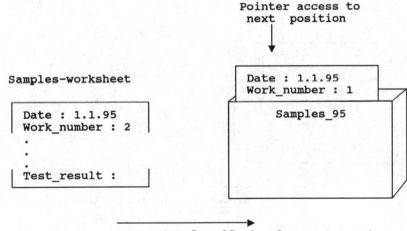

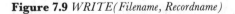

Figure 7.9 *WRITE(Filename, Recordname)*

empty, the file pointer will have been set to zero. The first instruction to WRITE(Filename, Recordname) will therefore place the first record in the file (Figure 7.8). All subsequent instructions to WRITE(Filename, Recordname) will place the records in the file sequentially, incrementing the file pointer each time (Figure 7.9).

7.5.3 Reading Records from a File

Again, reading records from a file is easy. The statement READ(Filename, Recordname) stores the record that is currently being pointed to by the file pointer in the variable name Recordname. The file pointer is then incremented, thus giving access to the next record. It is only possible to read from a file that has

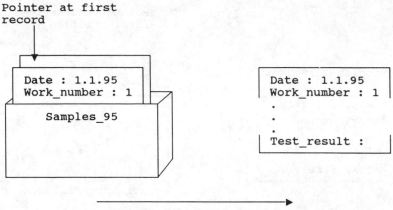

Pointer at first
record

Date : 1.1.95
Work_number : 1

Samples_95

Date : 1.1.95
Work_number : 1
.
.
.
Test_result :

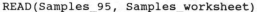

READ(Samples_95, Samples_worksheet)

Figure 7.10 *Reading the first record*

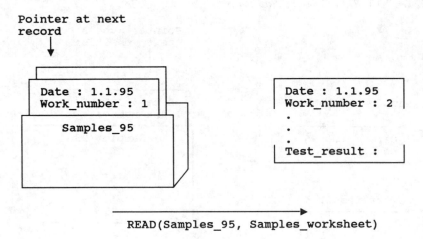

Pointer at next
record

Date : 1.1.95
Work_number : 1

Samples_95

Date : 1.1.95
Work_number : 2
.
.
.
Test_result :

READ(Samples_95, Samples_worksheet)

Figure 7.11 *After reading the first record*

records in it. Assuming therefore that our file contains records, we have operated the RESET command and not previously read from the file, the pointer will be at the first record (Figure 7.10). After execution of the first READ(Filename, Recordname), the pointer will automatically be incremented to give access to the next record in sequence (Figure 7.11).

7.5.4 Detecting End of File

There are two methods for detecting the end of the file. A flag will automatically become TRUE when, after successive READ(Filename, Recordname) steps, the file pointer has gone past the last record in the file. Alternatively, a 'dummy' record with a very high value in one of the data fields can be placed at the end of the records (Figure 7.12).

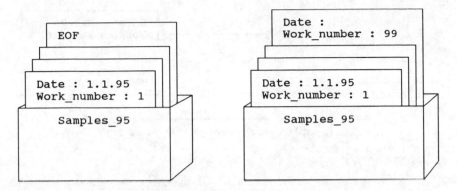

Figure 7.12 *End of file markers*

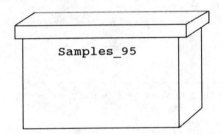

Figure 7.13 *Closing a file*

7.5.5 Closing a File

Closing a file is done for you in Pascal! This, however, is not the case with other languages (Figure 7.13).

7.5.6 Record Manipulation

It is possible to access individual fields with the READ and WRITE commands, READ(Recordname.Fieldname), and WRITE(Recordname.Fieldname).

In our case we could have, READ(Samples__worksheet.Work__number). This would allow us to read the Work__number. To read the next field, another

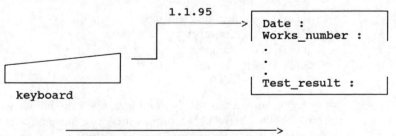

Figure 7.14 *WRITE(Samples__worksheet. Date)*

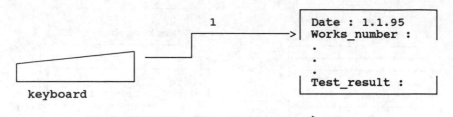

WRITE(Samples_worksheet,Works_number)

Figure 7.15 *WRITE(Samples__worksheet. works__number)*

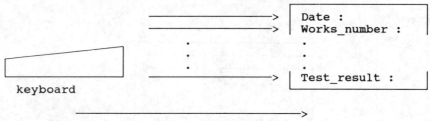

WITH Samples_worksheet

Figure 7.16 *The WITH command*

statement is needed, for example, READ(Samples__worksheet.Customer__num-ber), and so on for all the fields in the record. The same applies to the WRITE(Recordname.Fieldname) statement (Figures 7.14 and 7.15). In the case of records with large numbers of fields, or just for convenience, it is possible to use the WITH statement (Figure 7.16).

```
WITH  Samples__worksheet DO;
      BEGIN
              READ(Date);
              READ(Work__number);
              READ(Customer__number);
              READ(Customer__name);
              READ(Customer__address);
              READ(Sample__number);
              READ(Sample__material);
              READ(Test__number);
              READ(Test__name);
              READ(Test__result);
      END;
```

The algorithm for generating a new file is as follows: open the file for writing and initialize the variable choice to 'Y'; as long as there are more records to be entered, read in the record details and write them to the file; and close the file. This can be represented on a flowchart (Figure 7.17). In effect then, we would have a suite of such programs for each file. Should you wish to read more on this

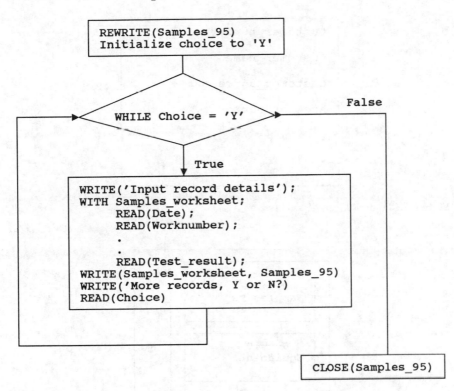

```
REWRITE(Samples_95)
Initialize choice to 'Y'
```

```
WHILE Choice = 'Y'
```
False

True

```
WRITE('Input record details');
WITH Samples_worksheet;
     READ(Date);
     READ(Worknumber);
        .
        .
     READ(Test_result);
WRITE(Samples_worksheet, Samples_95)
WRITE('More records, Y or N?)
READ(Choice)
```

```
CLOSE(Samples_95)
```

Figure 7.17 *Algorithm to generate a file*

subject I can strongly recommend 'Language Independent Design—An Introduction', NCC/Blackwell.

7.6 PROBLEMS WITH TRADITIONAL FILE-BASED DATA PROCESSING

Let us now expand our collection of files to include a file that holds customer details as shown in Figure 7.18. Our filing system now looks something like Figure 7.19. Each file contains data that is common to other files. Let us examine the consequences of this.

Duplication and Redundancy. There are three data fields common to the files Samples—worksheet and Customer—details, namely Customer—number, Customer—name, and Customer—address. Data are duplicated (Figure 7.20). For very large volumes of data this could mean that memory is wasted on storing duplicated data.

Chronological Inconsistency. The fact that the data are possibly being held in many different files leads to other problems. If data are updated in one file in the system and not in the others, inconsistency has been introduced. Consider if the customer changes address. It is necessary to amend each file accordingly, which

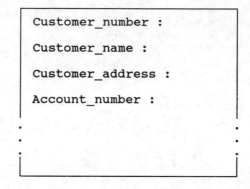

Figure 7.18 *Customer__details record*

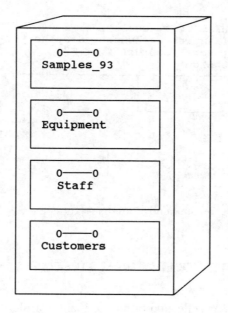

Figure 7.19 *Laboratory file system*

takes time. It is possible that at a given time only one file has been amended and the other remains to be changed. Interrogation of the files at that point in time will give different results (Figure 7.21). At a later date, when both files have been amended, there will be consistency. The inconsistency depends on when the files are interrogated—*chronological inconsistency*. This problem is compounded if the common data are more geographically dispersed. No doubt the accounts department, in a different building, also holds the customer address (Figure 7.22).

Lack of Integration and Control. This is an inevitable consequence of the above. If the data are replicated across a number of files in varying format, it is very difficult to have effective over-all control and monitoring.

Samples_worksheet   Customer_details

```
┌─────────────────────────┐   ┌──────────────────────────┐
│ Date :                   │   │ Customer_number :        │
│ Work_number :            │   │ Customer_name :          │
│ Customer_number :        │   │ Customer_address :       │
│ Customer_name :          │   │ Account_number :         │
│ Customer_address :       │   │                      .   . │
│ Sample_number :          │   │   .                  .   │
│ Sample_material :        │   │   .                  .   │
│ Test_Number :            │   │   .                  .   │
│ Test_Name :              │   └──────────────────────────┘
│ Test_result :            │
│ Analyst :                │      File Customers
└─────────────────────────┘
```

        File Samples_93

Figure 7.20 *Duplication*

Samples_worksheet   Customer_details

```
┌─────────────────────────┐   ┌──────────────────────────┐
│ Date :                   │   │ Customer_number :        │
│ Work_number :            │   │ Customer_name :          │
│ Customer_number :        │   │ Customer_address :       │
│ Customer_name :          │   │ old address              │
│ Customer_address :       │   │ Copenhagen               │
│ Vesterbrogade            │   │ Account_number :         │
│ Copenhagen               │   │                      .   │
│ Sample_number :          │   │   .                  .   │
│ Sample_material :        │   │   .                  .   │
│ Test_Number :            │   │   .                  .   │
│ Test_Name :              │   └──────────────────────────┘
│ Test_result :            │
│ Analyst :                │      File Customers
└─────────────────────────┘
```

        File Samples_93

Figure 7.21 *Chronological inconsistency*

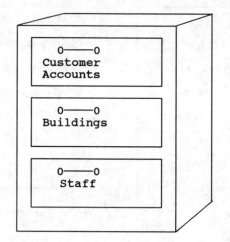

Figure 7.22 *Accounts file system*

Data Accessibility. Geographically dispersed data are harder to access. This means a trip to the accounts department!

Data Privacy/Confidentiality. One solution to the other problems is to have a record format that holds all the information, but at one centralized site only, *e.g.*, Samples_worksheet holds data fields previously held on Customer_details, *etc.* The problem is that the records become extremely large and it may be that some of the data fields contain confidential information that is not relevant to all the applications. Data security should be on a 'need to know basis'. Furthermore, data access is a problem.

To summarize, decentralized files are locally administered with security on a 'need to know basis'. There are common, not shared, data with update synchronization at intervals giving possible chronological inconsistency.

Characteristics: N files
N locations

Advantages	Disadvantages
small files	redundant data
security	inconsistent data
local control	common data
distributed	

A single, centralized file solves the problem of redundancy, but only at the expense of data security.

Characteristics: 1 file
1 location

Advantages	Disadvantages
no redundant data	very large file
no inconsistent data	no security
shared data	no local control
	not simply distributed

7.7 THE DATABASE CONCEPT

The traditional file-based approach is to define a file or set of files to support each application, and to write programs to process the data stored in these files. As we have seen this approach has major disadvantages including chronological inconsistency and redundant duplication. A *database* stores data common to all the applications, and the necessary *relationships* between the data items. The applications can then use the relevant data from this single source. With this approach, it is only necessary to record once data that are common to the applications (Figure 7.23). When the data are updated by one application, they will automatically be updated for all other applications. By doing this, we avoid redundancy and achieve consistency. Also, integration of data between the different applications is automatic, and central control is achievable. Each application has its own 'view of the data', which is defined by the database management system (DBMS). A database can be defined as: *a collection of related, non-redundant data shared by different applications on a need to know basis.*

Databases provide data consistency, concurrency, accessibility, privacy/confidentiality, and transparency.

Data Consistency. There are no redundant data, therefore there is no chronological inconsistency.

Data Concurrency. In systems where there are several users handling the data at the same time, the database ensures that there is no confusion if two or more users try to read from or write to the same record simultaneously. There is centralized data control.

Data Accessibility. Each data item is available to all those users who need it.

Data Privacy/Confidentiality. Data are protected from access and change by unauthorized users.

Data Transparency. To simplify the work of users, they do not need to know where the data is stored, how the data is stored, or what other data are in the system.

A system that controls the use of the data in a way that satisfies these requirements is known as a DBMS. Over the years, a number of different approaches to implementing DBMSs have been taken. The most common approaches have been the hierarchical, network, and relational databases.

Let us now consider the advantages and disadvantages of each type of database structure. In our example, each sample has a unique Sample_number and is typically divided into aliquots. A different test is performed on each aliquot. Each test is performed by an analyst on a named and numbered instrument.

```
Sample_number      :
Test_number        :
Test_name          :
Analyst_number     :
Analyst_name       :
Instrument_number :
Instrument_name    :
Result             :
```

Application 1
view

Application 2
view

Customer_details   Global data   Samples_worksheet

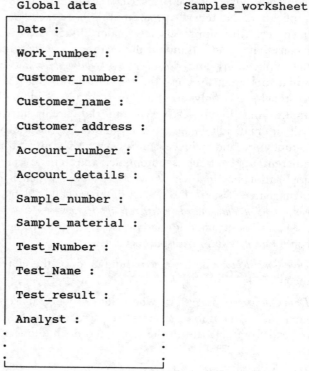

```
Date :

Work_number :

Customer_number :

Customer_name :

Customer_address :

Account_number :

Account_details :

Sample_number :

Sample_material :

Test_Number :

Test_Name :

Test_result :

Analyst :
```

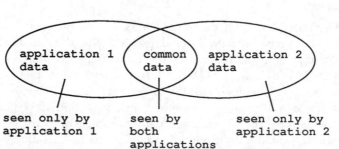

Figure 7.23 *The database*

7.7.1 Hierarchical Databases

One of the early examples of a DBMS was that introduced by IBM in 1968. It was called the Information Management System (IMS) and it is still used today on mainframes. In this approach, the data are viewed *hierarchically*. This means that sets of records containing data are related to each other by an ownership relationship, *i.e.*, one set of records owns another set of records. This can also be

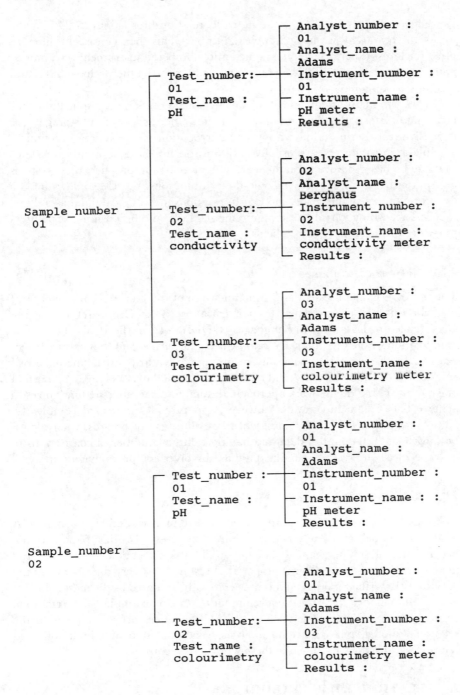

Figure 7.24 *Hierarchical database*

viewed as a tree structure, *i.e.*, parent–child relationships (Figure 7.24). At the top of the hierarchy we have Sample_number. This then consists of different tests. Each test is performed on a named and numbered instrument by a named and numbered analyst. The data are organized on a strict hierarchical basis, and a child may only have one parent.

One of the main disadvantages of this organization is data duplication. For large data systems this duplication is inefficient and, as a result of storage cost, expensive. Furthermore, navigation or interrogation of the system is complex. The hierarchical database is usually only suitable for those information systems that can be based naturally on the hierarchical model; in practice, these systems are quite rare. In order to overcome these limitations, the IMS allows for the definition of logical databases and logical relationships. Logical databases shape the trees according to real-world requirements. Logical relationships use pointers to reduce duplication of record fields.

7.7.2 Network Databases

The network model is essentially a modification of the hierarchical model. The idea for this model originated from a Database Task Group set up by the Conference On Data System Languages (CODASYL) in the 1970s. The major modification to the hierarchical database is in the *relationships* that exist between records. In the network model, any record may own any other type of record (Figure 7.25). In this parent–child relationship, a child may belong to more than one parent. Using the network approach, data analysis removes redundancies to achieve better data integration; however, this provides increased complexity. Network databases typically consist of large numbers of records, each record holding a small amount of data and having a larger number of connections to other records. Interrogation of the database involves complex navigation.

7.7.3 Relational Databases

Relational databases were first introduced in 1970 by E.F. Codd in the paper: 'A relational model of data for large shared systems'. In this paper, which became the most important document on database technology, the *relational* model for database systems was proposed. This model is very different from the previously described models, and is now generally accepted as the most coherent and usable model for DBMS development. As an example of a relational database, we will make reference to ORACLE. There are three parts to the relational model, relational data analysis, relational data manipulation, and relational data integrity. Each of these will now be considered.

7.8 RELATIONAL DATA STRUCTURES

It is possible to form a series of tables for all the data in a given system. These tables would be unnormalized because they would contain repeating groups of fields, ambiguities, and data redundancies. The technique of relational data

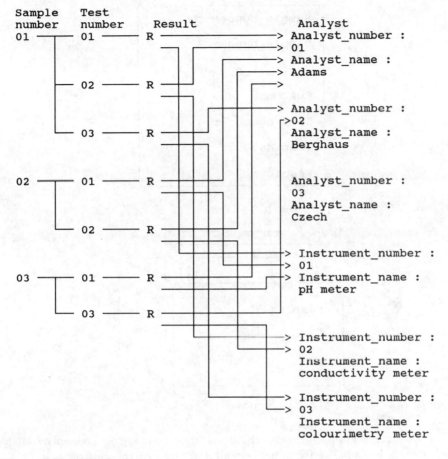

Figure 7.25 *Network database*

analysis addresses the following: complete and detailed definition of data; identification of data interdependencies; resolution of all ambiguities; elimination of unnecessary data duplication; creation of a set of relations, each having a unique key, and of data that are totally dependent on that key.

In a relational database, data are stored in files that are represented by two-dimensional tables called *relations*. The operations used on these tables are mathematically founded in *relational algebra*. As an example, consider the simplified record we used previously, Samples__worksheet, which was held in the file called Samples__93 (Figures 7.26 and 7.27). The records in this field can be rearranged in two-dimensional table form, with each Sample__number representing a record as shown in Figure 7.28. Our primary key, Sample__number, is underscored. This relation (table), called Samples__worksheet, has seven *attributes* (*columns*) and eight *tuples* (*rows*). Attributes are the data fields of records, tuples are the data values for a record (Figure 7.29).

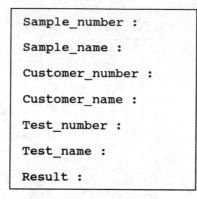

Figure 7.26 *Simplified record—Samples__worksheet*

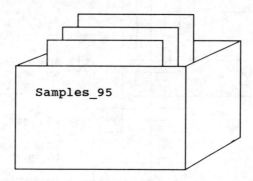

Figure 7.27 *File of simplified records Sample__95*

A relation must fulfil the following rules. If these rules are obeyed, only a small set of operations are needed to perform all data retrieval demands.

(i)   *Each relation has a name.* Every table has a name; therefore a table can be referred to by name. In our case the name is Samples__worksheet.

(ii)   *No two tuples of a relation can be the same.* A table cannot contain two identical rows. Each row must have a column/attribute that can be used as a key. A key is a field or group of fields that can be used to uniquely identify tuples in the table. A key that consists of a single field is called a simple key. A compound key is a group of fields used to identify uniquely tuples in a table. For example, there is no single field capable of providing a unique key for our relation as Sample__number is associated with many Test__numbers. However, a combination of Sample__number and Test__number will provide a unique compound key.

(iii)   *The order in which the tuples are presented is not important.*

(iv)   *Each attribute has a unique name.* This means that it is possible to refer to a specific column.

(v)   *The order of the attributes is significant.* This means that two relations containing the same attributes, but in a different order, are, strictly speaking, different. However, the order of the attributes is not significant in most practical situations.

Sample_number	Sample_name	Customer_number
01	oil	01
02	finds	02
03	oil	01

Customer_name	Test_number	Test_name	Result
	01 02 03	pH cond col	7 1.2 2.3
	01 02	pH cond	6 3.4
	01 02 03	pH cond col	5 4.5 5.5

Figure 7.28 *Samples__worksheet*

(vi)   *Any tuple/attribute intersection is a single value.* Each attribute in a tuple is said to be atomic. This concept is important, because it enables a small, but powerful, set of operators that satisfy all the standard data retrieval needs of a DBMS, to be defined for relational systems.

(vii)   *The values in any one column are all from the same domain.* A domain is a set of allowable values. It is important to know the domain for each attribute. This allows the programmer and the DBMS to determine which variables are valid and which are invalid.

(viii) *Each tuple is uniquely identified by a single attribute or a combination of attributes.* This is known as the key of that tuple.

(ix)   *The primary key must not be null.* If the primary key were null it would be impossible to identify the tuple. This rule is known as the entity integrity constraint.

(x)   *If a value in one relation refers to a tuple in another relation, this second tuple must exist.* This rule is known as the referential integrity constraint.

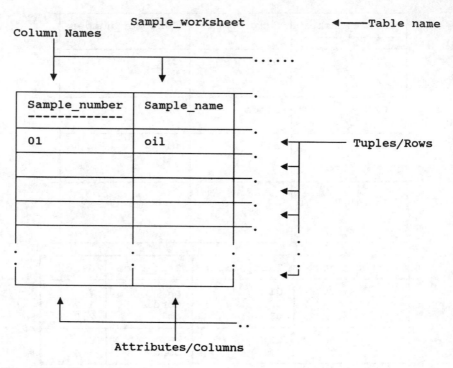

Figure 7.29 *Tuples and attributes*

7.9 DATA ANALYSIS

In the early stages of designing a computer system, the analyst(s) will carry out an investigation of the needs of the user and user groups within the organization. Based on the results from this investigation, the analyst(s) will be able to decide which data should be held in the database. This decision is based on the needs of the users, *i.e.*, which reports have to be written, which enquiries do the users need to make, which data will be available to which users, *etc*. The designer will then combine this information to design a database meeting the requirements of the users.

7.9.1 Data Normalization

When it is decided what data are to be held in the database, the designer faces the problem of how to store this data in the most efficient manner. For example, it is the responsibility of the designer to remove duplication of non-key data. The general goal of *data normalization* is to produce a set of entities that will support the users' needs and will conform to the requirements of the database. The reports and documents handled by the users form the starting point for this normalization. For each of these documents, a *local* data model is developed. When this is achieved, the various subschemes are combined and consolidated to form a *global* data model or schema. This is the final global data model that forms the basis of the database.

There are three standard steps that are followed to create a suitable set of relations. These steps are called the *first, second,* and *third normalized forms*. There is some academic discussion about fourth, fifth, and even higher normalized forms, but the third normalized form is usually adequate for any commercial DBMS. To be able to explain the three steps in a little more detail some basic concepts must first be defined.

Functional dependency. The following is a definition of functional dependency: 'an attribute B is functionally dependent on another attribute A (or possibly a collection of attributes), if a value for A determines a single value for B'. Put simply, if there is a value for A, can we find a single value for B? If there is, B is functionally dependent on A. Using our relation Samples_worksheet, supplying a Sample_number will give three values with the attribute Test_number. The attribute Test_number is not functionally dependent on the attribute Sample_number.

Primary key. The definition of a (primary) key is based on the concept of functional dependency and is as follows: 'attribute A (or a collection of attri-

Sample_number	Sample_name	Customer_number
01	oil	01
02	finds	02
03	oil	01

Customer_name	Test_number	Test_name	Result
	01	pH	7
	02	cond	1.2
	03	col	2.3
	01	pH	6
	02	cond	3.4
	01	pH	5
	02	cond	4.5
	03	col	5.5

Figure 7.30 *Samples_worksheet, un-normalized*

Sample_number	Sample_name	Customer_number	Customer_name
01	oil	01	
02	finds	02	
03	oil	01	
⋮	⋮	⋮	⋮

Figure 7.31 *Samples—worksheet, first normal form*

butes) is the primary key for a relation R, if (i) all attributes in R are functionally dependent on A, and (ii) no subset of the attributes in A (assuming A is a collection of attributes and not just a single attribute) also has property 1'. In those (rare) cases where more than one attribute can serve as a primary key, one of the attributes is designated to be the primary key and the other one(s) is(are) referred to as candidate key(s). We have now defined the basic concepts and can start discussing the actual normalization process.

7.9.1.1 Unnormalized Form. We will illustrate the normalization process for our relation Samples—worksheet (Figure 7.30). The unnormalized data form is the series of records that we have put in tabulated form. The primary key for the relation is Sample—number (underlined).

7.9.1.2 First Normalized Form. The first normalized form involves removing repeating groups from a relation. If we look back at the formal properties of a relation, we see that rule (vi) says that at any attribute/tuple intersection there is a single value. However, in the real world it is often the case that certain pieces of

Sample_number	Test_number	Test_name	Result
01	01	pH	7
01	02	cond	1.2
01	03	col	2.3
02	01	pH	6
02	02	cond	3.4
03	01	pH	5
03	02	cond	4.5
03	03	col	5.5
⋮	⋮	⋮	⋮

Figure 7.32 *Sample—results, first normal form*

data are repeated. In our example, there are several tests associated with each sample. The sample is divided into aliquots and a test is performed on each separate aliquot. Before the database is implemented, these repeating groups have to be removed (Figure 7.31). To do this, we create a new relation that contains the attribute(s) containing the repeating groups. In this new relation, the repeating groups are resolved into tuples with atomic values. The primary key for this relation is a combination of the original primary key and another attribute, Sample__number, and Test__number (Figure 7.32). We see that the repeating groups from the original relation have been resolved into a new relation (table), the key for which is the composite key made by the attributes Sample__number and Test__number. The original key has been extended. A 'relation' that contains repeating groups is, strictly speaking, not a relation, according to the relation rules. When the repeating groups have been removed, the relations are said to be in first normalized form.

7.9.1.3 Second Normalized Form. For a relation to be in second normalized form implies that if the relation has a composite primary key (a primary key composed of more than one attribute), then any attribute in the relation must be functionally dependent on the entire primary key and not just a part of the primary key. Also, the relation must be in first normalized form. Attributes that are only functionally dependent on part of the key should be removed to form a new relation. The primary key for the new relation would then be that part of the original primary key on which the attribute is dependent. Relations that are in first normalized form and have a single attribute as primary key are, by default, in second normalized form. The reason for reducing the data to second normalized form is to avoid data duplication. If an attribute is only dependent on a part of the primary key, the same data may be found in many tuples. Consider the relation, Sample__results. In this relation, we see that the attribute Test__name is only functionally dependent on a part of the key, namely on the Test__number. Therefore, data are duplicated across several tuples. The Result attribute is functionally dependent on the entire composition key. To bring the relation to second normalized form, we construct a new relation, Tests (Figure 7.33, 7.34, and 7.35). Data duplication can (apart from it being a waste of memory) lead to inconsistency problems. We have now reduced the data to second normalized form.

Sample_number	Sample_name	Customer_number	Customer_name
01	oil	01	
02	finds	02	
03	oil	01	
.	.	.	.
.	.	.	.

Figure 7.33 *Samples__worksheet, second normal form*

Sample_number	Test_number	Result
01	01	7
01	02	1.2
01	03	2.3
02	01	6
02	02	3.4
03	01	5
03	02	4.5
03	03	5.5
⋮	⋮	⋮

Figure 7.34 *Sample__results, second normal form*

Test_number	Test_name
01	pH
02	cond
03	col

Figure 7.35 *Tests, second normal form*

7.9.1.4 Third Normalized Form. When the relations have been converted to second normalized form, they are changed to third normalized form by removing any attributes that are dependent upon attributes other than the primary key, and by putting these in a separate relation. When the relation is in third normalized form, the attributes will be functionally dependent on the primary key and on

Sample_number	Sample_name	Customer_number
01	oil	01
02	finds	02
03	oil	01
⋮	⋮	⋮

Figure 7.36 *Samples__worksheet, third normal form*

Customer_number	Customer_name
01	
02	
03	

Figure 7.37 *Customers, third normal form*

Sample_number	Test_number	Result
01	01	7
01	02	1.2
01	03	2.3
02	01	6
02	02	3.4
03	01	5
03	02	4.5
03	03	5.5

Figure 7.38 *Samples—results, third normal form*

Test_number	Test_name
01	pH
02	cond
03	col

Figure 7.39 *Tests, third normal form*

nothing else (Figure 7.36, 7.37, 7.38, and 7.39). Again, the purpose of bringing the relation to third normalized form is to prevent data duplication.

Second normalized form relations that have attributes functionally dependent on the primary key are, by default, in third normalized form. The normalization process must be carried out on all the entities in the organization. When this is done, a global model can be formed.

7.9.2 Deriving the Global Model

We are now at the stage of having derived a number of local data models, one for each user view. However, the relations/entities do not exist in isolation, but have to be merged together to form a global data model that can be used when mapping to the physical DBMS. For example, a number of the derived relations may have the same key and could therefore be consolidated into one single relation. It is, of course, necessary to control user access to the attributes of this single relation, so that each user has access to the needed attributes, but to no other. When the global model has been derived, it is the responsibility of the database designer and database administrator to decide how the physical data are to be organized. It may, for example, be desirable to 'split' the global model again if some data in the global model are much more frequently used than other data. Furthermore, restrictions on how the data can be held may be imposed by the actual DBMS. When the model of the system has been derived, this has to be mapped to the physical DBMS. If we are using the relational model, for example ORACLE, we have to create relations, and implement the keys, nulls, and other restrictions. All this will be done with the SQL Data Definition Language (DDL); however, this will not be discussed further.

7.9.3 Relational Data Manipulation

To extract data from the system, any DBMS must provide *data manipulation operations*. In a relational system only eight types of operation are needed to effect all possible combinations of data extraction and manipulation. These operations all take a set of tuples derived from one or more relations as their domain and map these to a new set of tuples that constitute a new relation. An example of this is the Select operation, which extracts specified tuples from a relation. The eight basic relational operations are Select, Times, Union, Minus, and Project, all of which are *primitive relational operations*, and Join, Intersect, and Divide, all of which are *derived relational operations*. To give an idea of the possibilities available, the functionalities of a few operations are discussed in more detail. The Select operation will extract one or more tuples from a relation in accordance with specified conditions to form a new relation. The Times operation yields the Cartesian product of two relations. The Join operation is a special, restricted form of this operation. The Union of two relations is the set of tuples that belong to either or to both of two compatible relations. When applied to two compatible sets of tuples, the Minus operation produces all those tuples belonging to the first set, but not to the second. The Project operation will extract all the values from an attribute or set of attributes in a relation with the resulting duplicate tuples being removed if necessary. The Join operation will combine relations, for example, on the basis of the two relations containing an identical attribute. Intersect operates on two compatible sets of tuples, and yields the tuples that belong to both sets. This operation can be derived from Union and Minus. When applied to two relations the attributes of which overlap, the Divide operation returns the non-overlapping attribute value for all those tuples in the first relation that match all the tuples in the second. Divide can be derived from Project, Select, Minus, and Join.

The use of these operations is best illustrated with an example. Here we will use SQL, which is the language used by ORACLE software.

7.9.4 Structured Query Languages

Data retrieval and manipulation is performed using *Query Languages* (or *Structured Query Languages*, SQLs). Structured Query Languages are not based on relational algebra, but the functionality of the listed relational operations does form a part of SQL. There are two parts to SQL, *SQL Data Definition Language* (DDL), and *SQL Data Manipulation Language* (DML). Strutured Query Language DDL is used for defining tables (*i.e.*, what columns does the table consist of, what are the data types, *etc.*), and SQL DML is used to insert, extract, update, and delete the data in tables created with DDL. Structured Query Language DML allows the database user to make enquiries about data in the database without having to write special programs. If we consider the ORACLE table Tests, a simple query could be

select Test_name from Tests where Test_number = 02

In answer to this query, ORACLE would respond

Test_name
_____

Conductivity

Structured Query Language provides many different facilities. If we were to want, for example, a table containing the functional descriptions and component codes from the Samples_tests table ordered by Test_number codes, we would write

select Test_number, Result from Samples_results order by Test_number

The ORACLE response to this would be

Test_number	Result
01	7
01	6
01	5
02	1.2
02	3.4
02	4.5
03	2.3
03	5.5

It is also possible to combine *where* and *order by*. For example, if we wanted to find the descriptions where the Test_number is greater than 01, and to show them in order by increasing Test_number we would write

select Test_number, Result from Samples_tests
where Test_number > 01
order by Test_number

The response to this would be:

Test_number	Result
02	1.2
02	3.4
02	4.5
03	2.3
03	5.5

As a *de facto* standard, SQL provides full functionality of the eight relational operations. To be relationally complete, a DBMS must support the basic data structures, and provide full functionality of the eight relational operations.

7.9.5 Relational Data Integrity

The ORACLE DBMS is relationally complete, but it is not fully relational. To be fully relational, a product must fulfil the following criteria: allow domain definition; provide entity integrity; provide referential integrity; and be relationally complete.

ORACLE provides a number of predefined domains such as integer, number, character, real, and data. The user is not allowed to define domains. *Entity integrity* requires that no tuple take on a null value for any attribute that participates in the primary key of a relation. If we consider the table Samples_test, the primary key is the combination of the attributes Sample_number and Test_number. Entity integrity then implies that all the tuples in this relation must have values in both the Sample_number and the Test_number attributes. Entity integrity ensures that it is possible to identify a unique tuple, and that no relation contains duplicate tuples.

Referential integrity requires that for any attribute or set of attributes in a relation that comprise a foreign key, the values assigned to these attributes must also be present in the relation for which they are the primary key. In the Samples_tests table, the Test_number column contains primary keys in the Tests table, so the values in the Test_number column contain foreign keys. Referential integrity does not allow values to be inserted in the Test_number in the Samples_tests tables if these values do not exist in the Tests table. Referential integrity would also restrict actions such as deleting values from the Test_number column in the Samples_tests table if these values existed in the Tests table.

As mentioned earlier, ORACLE uses SQL for defining and manipulating relations. Structured Query Language has no direct mechanism for defining or recognizing a key. It is possible to enforce entity integrity by requiring that certain attributes in a relation are not allowed to take on null values. As these do not have to be the primary key values, it is therefore possible to build relations in the system having duplicate tuples. This data structure is, strictly speaking, not a relation, therefore ORACLE 'relations' are usually referred to as 'tables'.

The ORACLE DBMS does not provide domain definition, and it does not automatically enforce referential and entity integrity for all relations. It can therefore not be described as fully relational. It should be mentioned that there are no fully relational products on the market. ORACLE is relationally complete, meaning that it provides the basic data structures and operations required by the relational model. The integrity rules are desirable, but not essential. To implement these rules would require significant extensions to the software without a corresponding increase in functionality.

7.9.6 Extensions to the Relational Model

The relational model covers issues such as data structures and data manipulation, but not important subjects such as the user interface, ease of data processing, and system performance. The ORACLE DBMS, for example, provides a number of enhancements to the relational model. A list of the related products follows.

SQL*Plus. Structured Query Language is the *de facto* standard language for building and querying relational databases. ORACLE has an extended programming interface called SQL*Plus. When used in an interactive mode, this provides a number of extra commands that can further process and format the output from an SQL command. It also provides facilities for editing and saving SQL command files. SQL is now an ANSI standard.

SQL*Forms. The SQL*Forms package allows the creation of form-based applications for manipulating and querying the database through customized screens.

SQL*Report. The SQL*Report allows formatting, specification, and generation of reports combining text with data derived and processed from an ORACLE system.

SQL*Menu. The SQL*Menu allows the construction of menu-driven applications that can call up reports, forms, and programs generated in other modules.

Easy*SQL. The Easy*SQL is an interface to ORACLE. It allows the user to create ORACLE applications with minimal keyboarding and programming.

SQL*Calc and SQL*Graph. The SQL*Calc and SQL*Graph are two modules that provide spreadsheet and graphics interfaces to ORACLE.

CASE (Computer Aided Systems Engineering). The CASE system is intended to provide support for the designers and implementors of computer based information systems. We will not investigate this further, but will mention that its goal is to provide a controlled environment for the design, implementation, and maintenance of complete systems.

7.10 REAL-TIME DATABASES

To be able to explain the concept of real-time databases, we must first recall the difference between real-time and interactive data processing. Real-time data processing requires that a certain specified response time is guaranteed, whereas this is not a 'hard' requirement in interactive data processing. With a real-time database, the time it takes to access the database and process the data in it must be deterministic and within a specified range. If, for example, we consider process control, such a database could be used for storing process values. Based on these values, a controller would then calculate the parameters for the actual process control. The controller therefore needs rapid and deterministic access to the process values stored in the database. Queries to a database are dependent on many different factors (*i.e.*, record size, number of records, organization of these, *etc.*) and the access time is therefore by nature non-deterministic. The development of a real-time database is therefore a very difficult task; I am not aware of any commercial database that could strictly be called a real-time database.

7.11 Other Relational Databases

Apart from the ORACLE relational database, there exist a number of commercially available databases based on the relational model: DB2 (from IBM), INGRES, UNIFY, RBASE 5000, RIM, RDB, INFORMIX, PARADOX, SYBASE, and FOXPRO, to mention but a few! They all share a basic functionality and support SQL as the standard language. Furthermore, they support the use of SQL embedded in third-generation languages such as C, COBOL, Pascal, *etc.* The main difference between third-generation languages like C and a language like SQL is that third-generation languages are procedural, whereas SQL is non-procedural. In a procedural language, the program's goal is accomplished with a series of instructions that specify in detail how this solution should be produced. In a non-procedural language, the program is a specification of the wanted result (for example, as in the SQL Select statement). The gain in user friendliness that is accomplished with a non-procedural language is often at the expense of efficiency. For example, typically, a non-procedural language does not allow loops, conditionals, *etc.* Version 6 of ORACLE provides a language PL/SQL, which gives support for SQL, but at the same time offers some of the functionality of third-generation languages, such as loops, conditionals, *etc.* Furthermore, ORACLE supports precompilers for SQL embedded in a large number of third-generation languages (depending on the working platform).

The basic facilities for application development provided by the different vendors are similar. Some of the products are extended with fourth generation development tools for GUIs. For example, ORACLE supports a product called ORACLE card. However, this product can only run on PCs (with Microsoft Windows) and Macintosh computers, whereas similar tools from some of the other vendors (for example, INFORMIX) are more 'general'. A strong point in favour of ORACLE is its ability to run on a number of different software (DOS, OS/2, UNIX, VMS, *etc.*) and hardware (mainframes, minicomputers, PCs) platforms. The ORACLE user is therefore not 'tied' to a few software/hardware platforms. Other products have more limited platforms. One such product is INFORMIX, which is limited to UNIX and DOS. This will no doubt change with time. In addition, ORACLE provides good networking facilities with support for a large number of industrial standard protocols. There is a difference in the target areas for the different products. Some products, for example INFORMIX, are targeted towards smaller/medium range applications. Other products such as SYBASE are targeted towards larger applications.

The main advantage of ORACLE is that it is a very mature and widely-used product that is portable to a number of different hardware/software platforms. However, ORACLE has a reputation for being slow and demanding when it comes to larger multi-user applications. The trend seems to be one of different vendors competing in the areas of on-line transaction processing (OLTP) and mission-critical applications. An analysis of the different database capabilities in these areas is beyond the scope of this text.

7.12 USING RELATIONAL DATABASES AND SQL

The relational model has many advantages when compared with other types of database. It is the only model that combines flexibility and simplicity. It also provides minimal redundancy and minimal inconsistency. Furthermore, the relational database is founded on mathematical principles. From the user's point of view, the main feature of the relational database is that it is very easy to use. Use of SQL gives the end user an efficient tool for database interrogation. The user can define format and content of the desired report in a flexible way, as permitted by the security controls. Although some query languages (such as ORACLE's SQL) allow data manipulation. SQL is usually only for data retrieval, not data maintenance. This must be considered as a disadvantage. Use of SQL can also place large demands on CPU time, if the user's query is malicious enough (for example, something like: 'Give me a list of all samples for the past 10 years').

7.13 Laboratory Information Management Systems

A Laboratory Information Management System (LIMS) is a database designed to meet the specific needs of laboratories. Enhancements to standard databases to meet laboratory requirements include direct instrumental data-logging, automatic instrument calibration monitoring, and process control. A wide range of LIMSs are available, varying in complexity from single-user operation to distributed multi-user systems, but they all integrate sample information with the results from instruments in order to reduce administration and increase throughput. As such they are an integral part of automated systems.

There are various ways to define a LIMS: a computerized system designed to provide on-line information about the analytical laboratory and the samples assayed within it; software to integrate the automation of instrumentation and data handling with the distribution of information within the over-all organization; and a computer-based system designed for the management of all laboratory resources (people and equipment) in order to collect, process, store, and integrate all laboratory data in an efficient, safe, and cost-effective manner.

Most LIMSs use a standard, commercially available, database platform thereby providing *de facto* standard SQL and a guaranteed upgrade path.

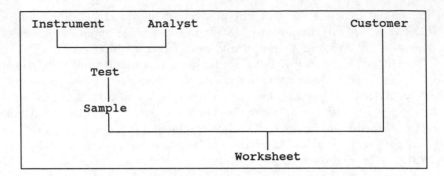

Figure 7.40 *LIMS structure*

Irrespective of the type of database used, the user is typically provided with a standard graphical user interface to allow ease of use. The platform is configurable to specific applications such as environmental chemistry, analytical laboratories, *etc.*, with minimum custom programming. Furthermore, there are seamless interfaces to and from third party software packages. In LIMSs the principal relationships have already been defined for the user (Figure 7.40). Some of the typical functionality found in LIMS will now be considered.

Laboratory Information Management Systems can be considered at two levels, management and operational. Prior to use, the data dictionaries should be configured; this is an initial management task. The LIMS may then be used for routine analytical tasks (Figure 7.41).

7.13.1 Configuration of LIMSs

Selection of the configure option provides the user with a pull-down menu from which a further selection may be made (Figures 7.42, and 7.43). The initial management task is password, instrument, analyst, and customer configuration.

7.13.1.1 Password Management. Data security requires password protection. This is typically on the hierarchical basis ranging from simple work registration to test and password management. Authority to access one level permits access to lower levels. Each user must have a unique ID code and a password that is known only to the user and the system manager. The authority level determines the LIMS function that a user can access. Any selected window option may be expanded to fill the complete screen (Figure 7.44).

7.13.1.2 Standard Controls. The entry of data and maintenance of the data dictionaries are achieved using a standard set of controls: Add (add record); Delete (delete record); Find, on the entry of a given field, the database is searched for a matching record; List, gives a listing of all the records, one of which may be selected; Update, allows record fields to be updated; Implement, causes the implementation of any changes; Exit, terminates the selected menu; and Help,

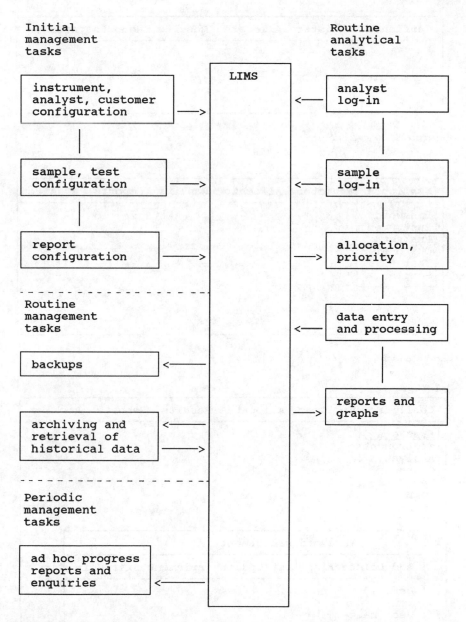

Figure 7.41 *LIMS operation*

provides user assistance. On selection of each configuration option, the associated window will be displayed. Each window will have the standard set of controls, but different data fields. Consider the instrument management window shown in Figure 7.45.

The instrument usage counter fields are automatically incremented and are used for calibration and service histories. On any counter attaining its reset

```
┌──────────────────────────────────────────────────────────────────┐
│ Configure   Register   Allocate   Results   Reports   Archive      │
├──────────────────────────────────────────────────────────────────┤
│                                                                    │
│ .                                                                .  │
│ .                                                                .  │
│                                                                    │
└──────────────────────────────────────────────────────────────────┘
```

Figure 7.42 *Initial LIMS screen*

```
┌──────────────────────────────────────────────────────────────────┐
│ //Configure//  Register   Allocate   Results   Reports   Archive   │
├──────────────┐                                                     │
│ Password     │                                                     │
│ Instrument   │                                                     │
│ Analyst      │                                                     │
│ Customer     │                                                     │
│ Test         │                                                     │
│ Sample       │                                                     │
└──────────────┘                                                     │
│ .                                                                .  │
│ .                                                                .  │
└──────────────────────────────────────────────────────────────────┘
```

Figure 7.43 *Selection of configure option*

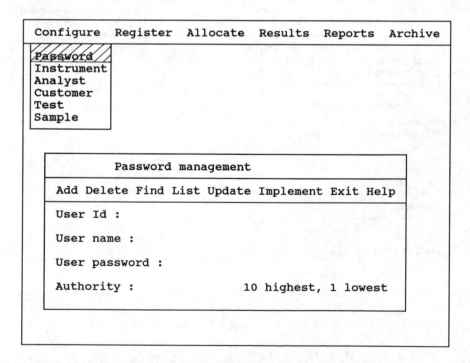

Figure 7.44 *Selection of password option from pull-down menu*

```
┌──────────────────────────────────────────────────────────────────────┐
│  Configure   Register   Allocate   Results   Reports   Archive         │
│ ┌─────────────┐                                                        │
│ │ Password    │                                                        │
│ │ Instrument  │                                                        │
│ │ Analyst     │                                                        │
│ │ Customer    │                                                        │
│ │ Test        │                                                        │
│ │ Sample      │                                                        │
│ └─────────────┘                                                        │
│                                                                        │
│     ┌────────────────────────────────────────────────────────────┐   │
│     │               Instrument management                         │   │
│     ├────────────────────────────────────────────────────────────┤   │
│     │ Add Delete Find List Update Implement Exit Help              │   │
│     ├────────────────────────────────────────────────────────────┤   │
│     │ Instrument Id :                                              │   │
│     │                                                              │   │
│     │ Instrument name :                                            │   │
│     │                                                              │   │
│     │ Instrument description :                                     │   │
│     │                                                              │   │
│     │ Calibration count :                                          │   │
│     │                                                              │   │
│     │ Service count :                                              │   │
│     │                                                              │   │
│     │ Total useage count :                                         │   │
│     └────────────────────────────────────────────────────────────┘   │
│                                                                        │
└──────────────────────────────────────────────────────────────────────┘
```

Figure 7.45 *Selection of instrument option from pull-down menu*

value, the LIMS offers the option to carry out the appropriate action. The decision of the user to accept or decline the request is automatically recorded. The counters are reset after the associated service/calibration (Figure 7.46).

Similarly, there are configuration windows for analyst and customer manage-

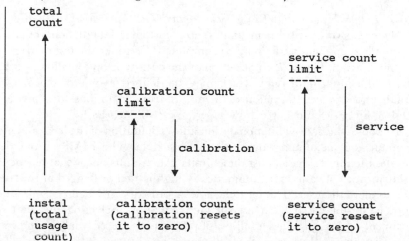

Figure 7.46 *Calibration and service usage counters*

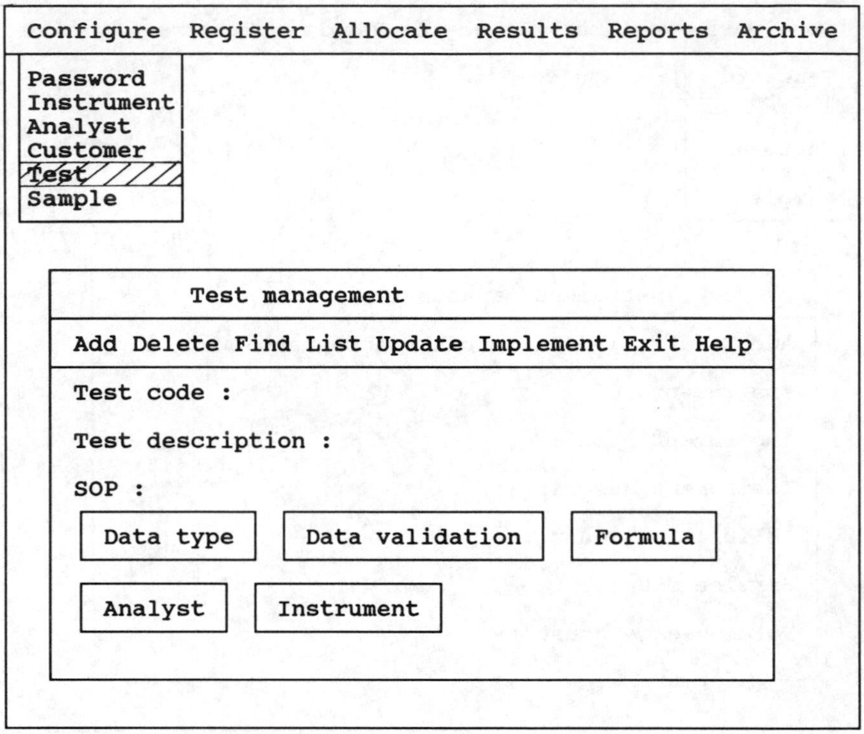

```
┌──────────────────────────────────────────────────────────────────────┐
│ Configure   Register   Allocate   Results   Reports   Archive          │
│ ┌─────────────┐                                                        │
│ │ Password    │                                                        │
│ │ Instrument  │                                                        │
│ │ Analyst     │                                                        │
│ │ Customer    │                                                        │
│ │ Test▨▨▨▨    │                                                        │
│ │ Sample      │                                                        │
│ └─────────────┘                                                        │
│                                                                        │
│    ┌──────────────────────────────────────────────────────────┐       │
│    │              Test management                              │       │
│    ├──────────────────────────────────────────────────────────┤       │
│    │ Add Delete Find List Update Implement Exit Help            │       │
│    ├──────────────────────────────────────────────────────────┤       │
│    │ Test code :                                                │       │
│    │                                                            │       │
│    │ Test description :                                         │       │
│    │                                                            │       │
│    │ SOP :                                                      │       │
│    │  ┌─────────────┐  ┌────────────────┐  ┌────────────┐       │       │
│    │  │ Data type   │  │ Data validation│  │ Formula    │       │       │
│    │  └─────────────┘  └────────────────┘  └────────────┘       │       │
│    │                                                            │       │
│    │  ┌─────────────┐  ┌────────────────┐                       │       │
│    │  │ Analyst     │  │ Instrument     │                       │       │
│    │  └─────────────┘  └────────────────┘                       │       │
│    │                                                            │       │
│    └──────────────────────────────────────────────────────────┘       │
│                                                                        │
└──────────────────────────────────────────────────────────────────────┘
```

Figure 7.47 *Selection of test option from pull-down menu*

ment. Each analyst may have a unique analyst code and an analyst name field. Typically, each customer record will have a variety of fields such as unique customer code and customer description field.

7.13.1.3 Initial Management Tasks—Sample and Test Management. Having completed the password, instrument, analyst, and customer data dictionaries, the test and sample data dictionaries may be completed. Consider the test management window shown in Figure 7.47. Selection of the data-type option allows configuration of the input data type, which may be defined as one of the following: textual, allowing free format text; menu, restricting options to a predefined selection; numeric; and single answer (*e.g.*, Yes/No, Pass/Fail).

Selection of the data-validation option allows definition of the local and global minimum and maximum values for numeric data. The LIMS automatically validates the input data against these limits. Any results outside either the local or global range of values are automatically highlighted or flagged at the time of entry to alert the user. The formula option provides the facility to produce a computed result based on a user-defined equation. Selecting the analyst or instrument options provides a full list of previously entered analysts and instruments from which to choose. Typically, only one instrument is associated with one test. We therefore establish relationships as defined in Figure 7.48.

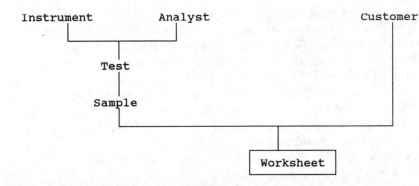

Figure 7.48 *Sample registration—the worksheet*

The sample management window typically provides the following fields: sample reference number; sample description; and sample tests.

It is possible to add or delete tests that are routinely carried out on a given sample whenever work is registered for that sample.

7.13.2 Routine Analytical Tasks

After the analyst has logged in, it is possible to perform sample log-in. Sample log-in is also called sample registration. In multi-user systems, the LIMS can be used at all terminals. The register option provides a window for sample registration. The associated sample number, date, and time can be entered manually or the LIMS will generate them automatically. Within this window it is possible to select a pop-up list of customers from which the appropriate one may be chosen. Similarly, there is a pop-up window for the selection of previously defined samples. The default tests associated with this sample are then displayed, *i.e.*, there is an automatically defined test regime (Figure 7.48). For current samples, the status of the tests are displayed. It is possible to add or delete tests. The facility exists for the batched entry of similar work, where most of the data input is common, and only needs entering once thus allowing the automatic assignment of default values for any input. Following the registration of work, the LIMS automatically produces labels for any samples. The labels display the reference number and date together with a choice of other information. Multiple copies of these labels are available for split-sample containers.

After sample registration, it is possible to perform resource allocation and prioritization. The LIMS schedules the work in any work area with regard to priority, submission date, and deadline date; this priority can be overridden. It is possible to display the work pending for a given instrument, analyst, customer or test.

7.13.2.1 Reports and Graphs. Using the reports option, it is possible to produce a variety of reports. During the initial report configuration it is possible to create reports according to the user specification, *i.e.*, titles, page headings and footings, line spacing, margins, column headings, data locations, *etc.* Some LIMSs have an integral word-processing facility. It is possible for the user to carry out some mathematical manipulation of the retrieved test data, such as the totalling or averaging of rows and columns, during the preparation of the report.

The LIMS automatically assigns a unique report reference number for every report produced, with an indication of the date. In accordance with GLP regulations, only one official copy of the final report is obtained from the LIMS. All other copies of reports for external use are marked 'PRELIMINARY' or 'DUPLICATE'. It is possible to print reports in either partial or complete form.

It is possible to perform detailed statistics such as regression analysis, chi-square, *etc.*, on any data in the LIMS. It is also possible to abstract relevant data into a file suitable for exporting to other programs, *e.g.*, LOTUS 123.

During routine operation of the LIMS we have log-in, testing, approval, and reporting. Log-in provides user-specific log-in screens, configurable pop-up windows for sample parameters, automatic sample IDs, an automatic date and time, and an automatic worklist. Testing consists of automatic data entry, manual data entry, automatic limit checking, and automatic calculations. Approval consists of instrumentation approval, and can be manual or automatic approval. Reporting provides user-configured reports, *ad hoc* reports, a seamless interface to third-party software, and automatic reporting.

7.13.2.2 Error Checking and Security. The menu structure interacts with the LIMS security so only functions and operations available to the user's authority class are accessible. Although the LIMS can display records in any file, access to specific information is limited to users belonging to the appropriate authorization level. Any additions, deletions, and revisions of the LIMS database, including dictionary files, are recorded automatically by an audit trail into an event log in accordance with GLP considerations, giving the action, date, time, and user identification.

7.13.3 *Ad hoc* Progress Reports and Enquiries

The ability to extract management information from the LIMS is a prime requirement. The LIMS is able to provide a rapid response to customer enquiries about the status of their work *i.e.*, sample tracking. The database can be interrogated on a wide variety of criteria. It is therefore possible to produce a full range of *ad hoc*, regular reports and statistics for monitoring the progress of work. Examples of the types of information that can be provided are: work scheduled for the day or week; work received on the day; work not completed by the deadline date; samples suitable for disposal or awaiting collection; work costed, but not invoiced per customer; work registered but not allocated to a work area; worksheets distributed, and to whom; work or analytical tests outstanding in order of priority and with respect to the deadline date; status of outstanding work and analytical tests, with identification of customer, priority, and relevant dates; all work, work per customer/name/code/abbreviation and work per customer type in a requested time period, specifying individual categories and work areas; all analytical tests, analytical tests per customer name/code abbreviation, and analytical tests per customer type in a requested time period, specifying indi-

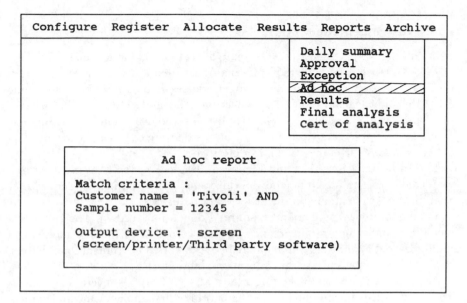

Figure 7.49 *Reports option*

vidual test parameters and work areas; any work/analytical tests out of specification, criticized, or unsatisfactory within a requested time period; quality assurance requests, findings and conclusions, in a requested time period, specifying individual test parameters and work areas; and cost evaluation of specified work/analytical tests within a requested time period, *i.e.*, quoted/estimated cost, calculated cost, and actual cost comparisons.

7.13.3.1 Ad hoc *Reporting*. *Ad hoc* reporting is interrogation of the database, usually associated with customer enquiries, and it can be in an unusual or unanticipated manner. Selection of the *ad hoc* option provides a window that allows the user to define search criteria (Figure 7.49). Let us assume that there is a telephone enquiry from the company Tivoli about the status of submitted sample. The sample number is 12345. The match criteria are completed, in this case using the logical connective AND. Other logical connectives include BETWEEN, NOT__BETWEEN, <, and >. The match criteria determine what information will be included in the report according to the defined criteria. The possible criteria may be selected from pop-up window menus. Alternatively, if supported, SQL search techniques may be used. It is possible to select extremely complex criteria.

7.13.3.2 Routine Management Tasks—Back-ups and Archiving. All data ready for archiving are catalogued. The data can be in database, word-processed or report format. The catalogue routine creates an index for every occurrence of every word across all data. Against this index, the LIMS stores the data, giving file name and storage identification. The LIMS is able to supply every reference whether in report, database or word-processed format, and to state where data are stored.

7.13.4 Instrumentation Interfacing

Different signals from instruments may be captured, for example, RS 232, IEEE 488, current loop, digital, and analogue. Identification by bar coding is possible. However, in a significant number of cases, half of the effort of installing a LIMS can be expended on interconnecting non-standard devices. Proprietary standards are the typical rule in laboratories—standardization is in its infancy. Consequently, there are various organizations that are concerned with interfacing standards; some of these are now described.

The Institute of Electrical and Electronic Engineers (IEEE) is involved in a variety of projects including the development of international standards for analytical instrumentation in the health-care industry.

The Laboratory Automation Standards Foundation (LASF). The LASF is a recently founded non-profit organization concerned with the development of standards that enable data interchange between varied instruments, computers, and application packages within automated laboratories.

The American Society of Testing Materials (ASTM) has several committees dedicated to instrumentation interfacing within both manufacturing and clinical applications. In particular, the ASTM is concerned with clinical laboratory systems, computerized manufacturing, LIMSs, instrumentation interfaces, *etc*.

The Consortium on Automated Analytical Laboratory Systems (CAALS), an undertaking of the National Institute of Standards and Technology (NIST) and the private sector, addresses standards in laboratory automation.

The Analytical Instrumentation Association (AIA) is a vendor-based organization that addresses standards for data interchange between analytical instruments. It is particularly active in the chromatography field.

The Standards Commands for Programmable Instrumentation Consortium (SCPIC) is a vendor-based organization that is concerned with harmonizing the software standards for instrumentation interfacing.

7.13.5 Conclusions

Without doubt, the LIMS is an aid to laboratory automation, and provides a cost-effective solution to the automation of administrative tasks and routine test procedures such as sample management, status, and analysis. They also provide facilities for workload management and report generation with the associated customer billing. Further, they permit on- or off-line data entry and acquisition with the associated instrument calibration histories. For standard instruments such as balances and pH meters, data logging directly into the database and the associated calibration are routine tasks. However, restrictions and complexities exist for the interfacing of some instrumentation to a LIMS. Laboratory Information Management Systems allow networking to defined international standards, for example, Ethernet (CSMA/CD bus, IEEE 802.3). One important consideration in respect of technology and LIMSs is 'future proofing'. The laboratory is only one part of the larger manufacturing complex. The emphasis should therefore be on the analysis and design of a corporate data architecture, a

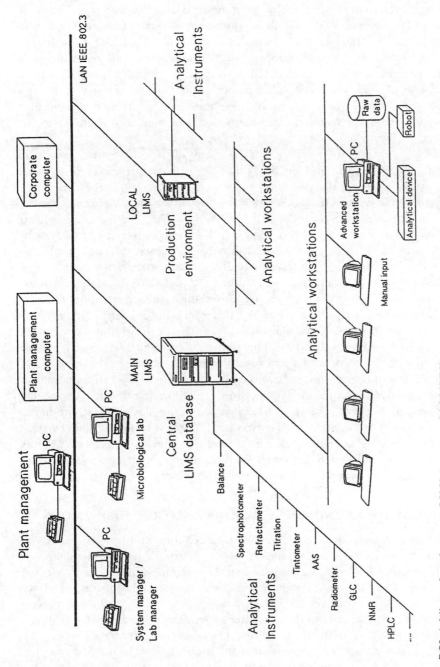

Figure 7.50 *A fully configured corporate LIMS (courtesey of COMPEX)*

LIMS being a subset of this architecture. Full consideration must therefore be given to corporate networking and the integration of databases to form a corporate data architecture thereby avoiding the dangers of having an 'island of automation' (Figure 7.50). This will be considered further in Chapter 9, Laboratory Information Management Systems. I like to think of the LIMS as a software tool that links the customer, samples, staff, instruments, and reporting in a controlled and regulated manner. In a properly designed automated LIMS laboratory all the necessary information is at the laboratory manager's 'fingertips' (Figure 7.51). The laboratory manager is therefore able to define and answer the questions: Who is doing what, when, where, and how? and Who did what, when, where and how?

Though real-time ability is claimed by many of the LIMSs this cannot be substantiated. The main problem being that 'real time' is often incorrectly taken to mean 'fast', without the associated guarantees of response time. The industry standard INGRES or ORACLE databases are often used. Operating system platforms include DOS for PCs with UNIX and VMS for workstations and minicomputers. The DOS itself is a single tasking, single user operating system. Attempts to make DOS real time must involve the timer, interrupt, and controller devices, and mechanisms to schedule tasks. Further, the problems of intertask communication, synchronization, and deadlock must be resolved. Questions must be asked about the real-time ability of the standard DOS or UNIX-based software packages. Further, Ethernet is not a deterministic networking protocol. The token-bus network, due to the deterministic nature of its token-based MAC method and its ability to prioritize frame transmission, is more suitable for process control. Laboratory Information Management Systems are restricted in handling graphics; however, the main criticism is certainly that, in themselves, LIMSs provide no guidance in the analysis and design of computer-based systems. Regardless of which LIMS package is employed, the environment under consideration must still be analysed, and the new computer-based system designed accordingly. The requirement is for a procedural framework that gives guidance for the full system life cycle, analysis, design, implementation, testing, and maintenance. In highly regulated environments this is essential.

7.14 FILE-PROCESSING APPLICATIONS AND CONSIDERATIONS

Dataprocessing systems are concerned with the capture and processing of data. Much data are in the form of transactions, *e.g.*, a sample needs to be analysed. This is sometimes called *transaction processing*. The way in which data are collected and processed gives rise to different processing modes or environments, which we will now consider (Figure 7.52). Users typically want immediate access to information. The designer of the computer system must decide whether this is a real need, as when staff are answering customer requests by access to the computer data file, or is an expression of the users' desires. In the first case, the system must be designed to handle continual and random enquiries that demand an index-sequential or direct-access file. With tape drives, a greater throughput of records is possible if records are processed sequentially. The design constraints

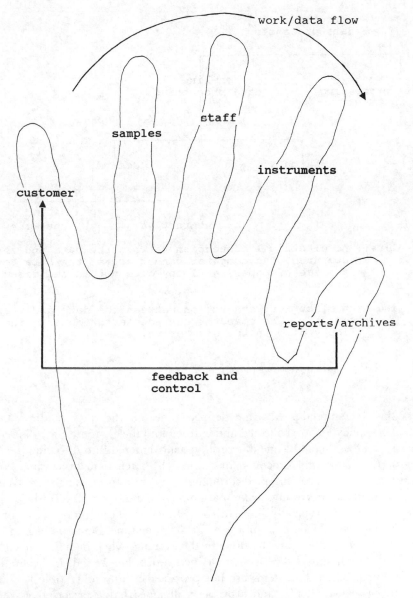

work/data flow

staff

samples

instruments

customer

reports/archives

feedback and
control

Figure 7.51 *LIMS functionality*

are linked to how much of each type of memory device depends upon the application. Different applications will have different requirements.

7.14.1 Batch Processing

With off-line data processing, the data are collected on index cards, pieces of paper or whatever. The important consideration is that the computer cannot

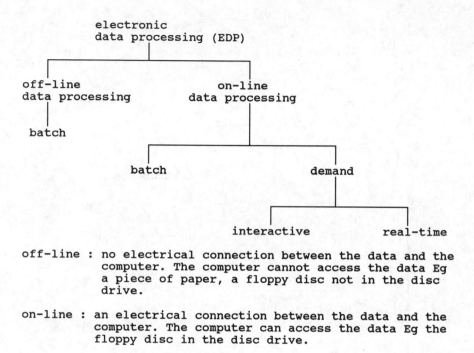

off-line : no electrical connection between the data and the
           computer. The computer cannot access the data Eg
           a piece of paper, a floppy disc not in the disc
           drive.

on-line : an electrical connection between the data and the
          computer. The computer can access the data Eg the
          floppy disc in the disc drive.

Figure 7.52 *Electronic data processing categories*

access this data directly. At some defined time, all the data relating to a
particular application are collected and entered into the computer; the data are
then said to be on-line. Off-line data are typically entered into the computer in
one group or batch and processed in one batch, batch data processing. The
identity of the group is maintained throughout the entire processing stage. With
this type of data processing, examination of the number of records to be
processed with time typically shows a ramp function. The number of records
builds up and then they are all processed at the same time. Batch processing on
mini- and main-frames is usually done in the evenings when demand for CPU
time is at a minimum. There is a predefined processing period, for example,
every evening. Batch data items are not processed at any other time (Figure
7.53). Using sequential files and a properly designed batch period gives very
efficient file processing. Certainly, one of the main advantages of batch processing is the ability to allocate resources. Furthermore, magnetic tape may be used.
However, there is the time delay between data capture and availability of the
processed data to users. Secondly, there is the problem of random enquiries. A
random enquiry would necessitate loading of the tape and then a sequential
search of all records preceding the desired one. In order to ensure efficient batch
processing, the batch period is such that there is a high hit rate, *i.e.*, most of the
records are processed.

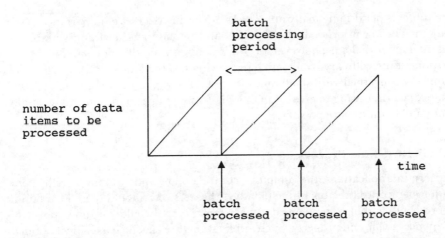

Figure 7.53 *Batch processing*

The off-line batch processing sequence is as follows: a batch is received; the data are keyed in (the transaction file); the input data are validated/revalidated; and the transaction file is sorted according to the master file to ensure a high hit rate.

Off-line batch data processing characteristics are as follows: the source documents are grouped into batches, input data are entered in groups with the identity of each group being maintained throughout the entire processing stage; typically concerned with small quantities of data, *e.g.*, one transaction per user; there is usually a considerable delay between data input and the availability of the results (the batch processing period); and enquiries are not possible.

With on-line data processing, there is a direct electrical connection between the data and the computer. The computer can, if necessary, control data input. The computer may, for example, perform a data logging function, say, reading the temperature from a reactor every 10 s. All these data are collected, but they can be processed at a later time, *i.e.*, batch on-line data processing.

7.4.2 Demand Processing

With on-line processing, the user can interact with the computer at any time via the I/O devices, which are under direct computer control. Therefore, data can be processed on demand. Unlike batch processing, there is no time lag between the data becoming available and the data being processed. With interactive processing, the data are processed as soon as possible. The response time of the computer system (software, hardware, number of users, *etc.*) can vary and the systems are classified accordingly. A typical computer user presses the return key in order to illicit a response from the computer. In a multi-user system, the computer is performing many other tasks of varying degrees of importance. Your task may be low priority. The computer will respond in a time frame suitable for an interactive dialogue; however, depending upon what other demands exist, the response time will vary. Interactive processing has the following characteristics: typically concerned with small quantities of data, *e.g.*, one transaction per user; immediate processing to give results; single transaction input; immediate validation; and enquiries are possible.

7.14.3 Real-time Processing

In certain applications, the computer controls plant and equipment. The computer may control the operation of valves, heaters, *etc.* Should there be a failure to respond within a predefined time frame, there would be disastrous consequences to life and property. Such applications are classified as critical applications. Real-time data processing is such that the computer must guarantee to respond to the data input within predefined time constraints. The correctness of real-time data processing depends not only on data, but on time as well.

Each application places different demands on the computer's memory requirements. Off-line batch data processing typically requires large data volumes with relatively relaxed performance demands. The emphasis here is high data volumes at low cost, an aim that can be met by using sequential files stored on magnetic tape or hard disk storage devices. At the other extreme, real-time data processing has relatively small data volumes; however, the memory device must be fast enough to meet the response-time requirements. These demands are met by IC memory devices.

7.14.4 Security: Confidentiality, Integrity, Availability

Data are valuable commodities, which, in addition, may be of a sensitive nature. As such there are potential threats to the security of that data. For example, if the data have commercial value, there may be attempts at fraud. Data can be copied very easily onto a floppy disk. Security is concerned with the protection of computer hardware, software, and data from accidental or malicious access, use, modification, destruction or disclosure. Security also pertains to personnel, data communications, and the physical protection of computer installations. In many applications, there must be guarantees that the data are correct, secure, and will always be available.

The normal data flow is from data source to data sink as shown in Figure 7.54. There are three main ways whereby data flow is insecure, by destruction, interception or modification.

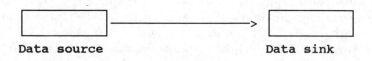

Data source   Data sink

Figure 7.54 *Normal data flow*

Figure 7.55 *Data flow interruption*

Destruction. A part or parts of the system (hardware or software) may be destroyed so that the data are no longer accessible. A temporary delay in ability to access data is called an *interruption* (Figure 7.55).

Interception. This is the unauthorized access to data that results in data being read and/or copied (Figure 7.56).

Modification. Unauthorized access also generates the risk of the data being modified or even new (incorrect) data being added (Figure 7.57)

The three main threats to security are confidentiality, integrity, and availability. I remember this by using the mnemonic CIA (Figure 7.58).

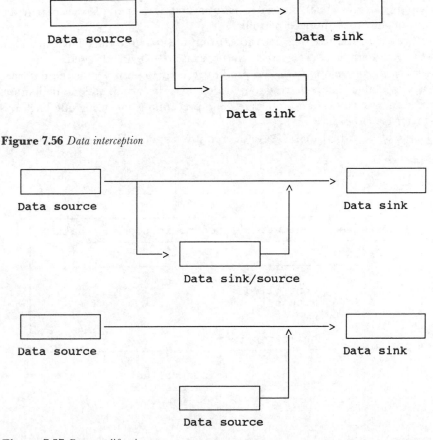

Figure 7.56 *Data interception*

Figure 7.57 *Data modification*

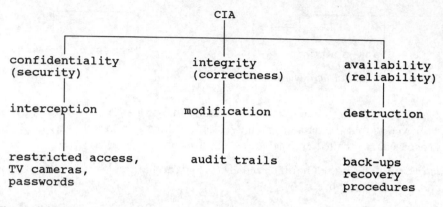

Figure 7.58 *CIA*

Confidentiality is concerned with data security. Security measures restrict the possibility of data interception or modification. Considerations include restricted access to computer equipment, passwords, and equipment monitoring, using television cameras. Such measures ensure that the data on the system are accessible only to authorized personnel.

Integrity is concerned with measures to ensure that the data have not been modified and are therefore correct. Audit trails are extensively used.

Availability, or reliability, is the procedure used to ensure that the data are always available. Data destruction, which may be accidental or deliberate, includes power failures and sabotage. The precautions taken include back-ups and recovery procedures.

The four assets that must be considered are, as shown in Figure 7.59, hard-

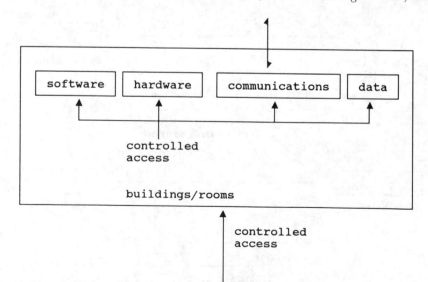

Figure 7.59 *Assets*

ware, software, data, and communications. Threats to these assets can be handled in the following ways: transfer, *i.e.*, insurance; reduction, *i.e.*, reducing the likelihood of threats; and retention, *i.e.*, estimating the cost of the loss (or not) and accepting the loss.

7.14.4.1 Security Layering. System security is concerned with restricting access to resources in a manner that is not prohibitatively restrictive to users. Typically there must be layers of security.

Access to the rooms/buildings must be controlled through, for example, visitor registration, escorting and use of an exit log. Access to equipment could be restricted simply by locking doors. Once the user accesses the equipment there should be keyboard locks. Constant monitoring is possible with television surveillance.

Protecting passwords by using a minimum of eight characters per password reduces the risk of the correct password being guessed or of programs being written to attempt password generation attempts. Passwords must not be displayed on the screen and access attempts should be logged with automatic cut-off after three erroneous attempts. Periodic review of the password log is recommended.

Passwords must be held on the computer in the password file. Successful access to the computer may give access to this file, which can be protected using encryption.

After successfully 'logging on' the users may be granted access to different sets of applications. This can be controlled by each user or user group having a user profile that defines permissible operations and file access rights. The access-control profile is implemented by an access matrix. One access of the matrix defines objects, the other subjects (Figure 7.60).

Modem use should be by authorization only. If the application demands, the use of dedicated or restricted telephone lines is recommended. There should be call-back verification.

7.14.4.2 Malicious Programs and Viruses. There are various malicious programs that can be introduced into a computer system. These vary in effect from a minor inconvenience to a complete system failure.

A Trojan horse is an apparently useful program or procedure that contains hidden code. Activation of the hidden code causes unwanted and typically deleterious effects.

User group	Applications	
	Application 1	**Application 2**
One	read, write, execute	
Two	execute	

Figure 7.60 *Access matrix*

A virus in this context is not a biological organism but a computer program that can modify or 'infect' other programs. This modification includes a copy of the virus program thereby allowing it to further infect other programs. Like biological viruses, computer viruses have a 'life cycle': insertion into the host computer system; a dormant phase while the virus waits for the activation trigger, which may be anything such as a time or a date; a propagation phase during which the virus makes copies of itself in other programs; and an execution phase during which the virus performs the functions it was designed to do (from simply clearing the screen to deleting all files!). There are many viruses including, for example, the Pakistani Brain that infects the DOS thereby infecting every floppy disk that is used.

As with all security measures, prevention is better than cure. All possible measures should be taken to prevent a virus from entering the computer system. On-going measures include program monitoring. The program monitor is triggered by system interrupts, which can then be examined for any unusual activity. Should a virus be detected, virus-removal programs may be used. Recovery is then the restoration of the computer system. On a lighter note, the following viruses are brought to the readers attention: US health care virus, it tests your system for a day, finds nothing wrong, and then sends you a large bill; Committee virus, it runs every program on the hard disk simultaneously, but does not allow the user to accomplish anything; and Star Trek virus, it invades your system in places where no virus has gone before.

7.14.4.3 Back-up Procedures and Recovery. Computer systems fail at both the hardware and software level. The methods that ensure a rapid restart of the system are called back-up and recovery procedures. These procedures provide for the recovery of data or software, for restarting the processing, or for the use of alternative equipment after a system failure or disaster.

The simplest form of back-up and recovery procedure is the generation technique: son, father, grandfather. Three generations of master file (and transaction files) are kept. Should master file 3 be lost or damaged, the system can always be reconstructed from master file 2 (Figure 7.61).

In on-line database systems the requirement to make regular copies of the disk and the time needed to do this may not be practical. The more commonly used methods of recovery in on-line database systems involve saving information about a specific transaction. Information about a record may be saved before the transaction is processed (the before image) or after processing takes place (the after image). The process of recovery may then be implemented in two possible ways.

With the before-image method, the database can be restored to its previous correct state if a failure occurs by using the before image. This is called roll back.

With the after image method, the database is copied at regular intervals and in the event of failure, recovery is achieved by taking the last copy and processing the after image records since that dump. This is called roll forward.

Disk shadowing is a technique whereby all updates to data on a main working disk(s) are duplicated (shadowed or mirrored) on a secondary disk thereby

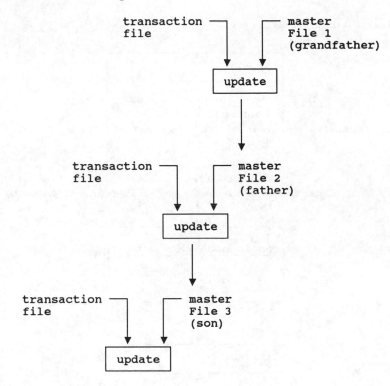

Figure 7.61 *File backups*

providing a fail-safe measure for critical data updates. In the final analysis, it MUST be possible to restore the system with minimal data loss in the shortest possible time. Most medium to large computer installations have documented back-up and recovery procedures.

7.14.5 Case Study

Union and Van den Bergh en Jurgens are two leading margarine manufacturers in Belgium and the Netherlands. The process consists of deodorizing (refining), and the addition of sodium hydroxide to remove fatty acids and of bleaching earth to remove colour and heavy metals. It is a batched process with a batch size of 6 tonnes. The process is controlled by microbiological, quality assurance and analytical laboratories.

Their process outline is as follows:

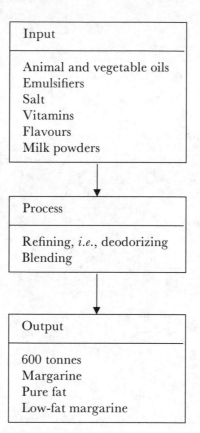

The analytical laboratory staffing is as follows:

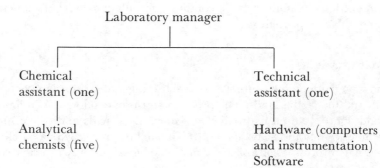

There are other personnel responsible for purchase, general administration, and cleaning. The laboratory is responsible for feedstock and finished goods. This requires a testing regime of approximately 500 samples per week, with about 10 tests per sample. The on-line tests include melting point, balances, pH, gas-liquid chromatograph (peak information, no graphics), nuclear magnetic resonance (solids, temp), free fatty acids, *i.e.*, test sequence of weight then titration, and texture analyser.

The LIMS used is UNILAB, a multi-user, multitasking package supplied by COMPEX. The relational database platform is INGRES 6.3, which runs on the UNIX (HP/UX 7003) operating system.

Hardware Configuration

(Hewlett-Packard) HP9000/835S
32 Mbyte RAM
2 × 300 Mbyte disks with IEEE 488 interfaces
2 × 600 Mbyte disks with fibre optic link interfaces
1 RuggedWriter (dot matrix printer)
1 Laserjet
1 HP7550 plotter (rarely used)
12 'dumb' terminals
5 PCs
1 modem
IEEE 802.3 LAN

It is worth noting that the 300 Mbyte disks are used only as spare disks and for storage of temporary files, and that the LAN is transparently connected to the company wide broadband network.

Recall that a single side of A4 paper is equivalent to 2K of binary (ASCII) data. We can therefore perform a 'filter-paper' calculation as follows:

500 samples per week
10 tests per sample

Text data	Computer data
assume 1 side of A4 text per test, *i.e.*, 5000 sides per week	assume 1 page = , *i.e.*, 10 000K = 10 Mbytes per week
50 weeks/year operational, *i.e.*, 250 000 pages/year	50 weeks per year operational, *i.e.*, 500 Mbytes per year

The system was fully implemented and operational with all necessary operational and security controls and also integrated to the production environment. An automated, 'future proofed' laboratory like this in a regulated environment is a beautiful thing to see.

CHAPTER 8

Software Engineering

The term 'software engineering' was introduced in the late 1960s in response to the 'software crisis' resulting from the introduction of third-generation computers based on IC technology. These machines increased, by an order of magnitude, the size and complexity of computer-based applications. Techniques applicable to smaller systems could not be scaled up resulting in overdue, unreliable, expensive, and difficult to maintain computer based data processing systems. The National Computing Centre, UK, commented that: 'Inadequate or incomplete analysis results in systems that do not satisfy user requirements. Poor design produces systems that are inefficient, ineffective, and inflexible. The consequences are increased cost, delayed implementation, error-prone operation and open-ended maintenance'. The quality of any computer system depends not only on the quality of the associated software and hardware but also on the operational procedures. In regulated industries, the quality of such systems must be demonstrated to be of an acceptable standard. The production of complex systems requires the use of the *software or system life cycle*. This consists of a series of distinct stages, each stage having clearly defined activities. The number and details of each stage vary, but, typically, after the initial statement of requirements there is (i) investigation of the current system—operations and problems; (ii) detailed specification of requirements and system specification; (iii) selection of technical options; (iv) system design; (v) coding, testing, and integration; (vi) implementation, testing, and system release; and (vii) maintenance.

8.1 METHOD REQUIREMENTS

A method is an organized approach to attaining a desired goal. For the analysis, design, and implementation of a large- to medium-sized computer-based system a method must be used. Most methods use the system/software life cycle and employ the basic principles of *stepwise, top-down decomposition*, which allows the deferment of detailed considerations by the use of abstraction to suppress and then progressively to emphasize detail, as appropriate. All methods aim to be understandable, expressive, implementation-independent, and generally applicable. Guided by a method, progression through the system development life cycle consists of a series of transformations from the user statement of requirements to the detailed design. This involves documentation employing different notations appropriate to the requirements of each stage. The statement of requirements document will be natural language with some graphics for clarity. From this document, the requirements-analysis stage has to produce a detailed requirements specification to be used as a reference document for all subsequent work and for final acceptance testing prior to handover. Progression through the development cycle typically reduces the natural language content with a subsequent increase in more diagrammatic notations. The output of each stage is a specification for the following stage from which the appropriate design is made.

The method used must have clear guide-lines, employing proven techniques and intrinsic documentation to fully identify, map, and validate the functions, events, and data. As such the method must provide a design that is *complete*, *consistent*, *correct*, and *unambiguous*. Completeness requires that all the necessary parts have been included. Completeness in design means that all the requirements have been met. Consistency is the property of logical coherence amongst constituent parts. Consistency may also be expressed as adherence to a given set of rules (FIPS Pub 101, June 1983). Intrinsic checks prescribed by the method are therefore needed. The method should also give guidance in maintenance procedures. In a highly regulated environment, use of a method is an invaluable aid to computer-system validation.

Method requirements include the following: (i) conformation to a recognized standard (national, international); (ii) comprehensive documentation with clearly defined stages and guide-lines; (iii) proven techniques that may be used as a communication aid to the end user and provide guide-lines for their use (*e.g.*, for the accurate identification of user needs and verification that these needs are being met); (iv) intrinsic documentation, *i.e.*, the method should produce system documentation as part of the analysis and design process; (v) consideration of security issues (access rights, audit trails, *etc.*); (vi) provision of quality assurance by defined procedures of verification and validation; (vii) consideration of maintenance issues; (viii) provision of a wide scope—strategic, tactical, and operational—thereby allowing computer-system development within the context of a 'corporate data architecture'; (ix) provision of a framework for the construction of computer related standard operating procedures (SOPs); (x) provision of quality assurance that conforms to GLP regulations, thereby providing a vehicle to laboratory accreditation; (xi) aid in the analysis and design of computer-based systems (LIMSs) with different degrees of automation, *i.e.*, manual data acquisition, automated data acquisition, and automatic data acquisition, where the application environment may include batched, real-time distributed systems; and (xii) techniques employed by the method should include those suitable for critical applications.

8.2 ANALYSIS AND DESIGN METHODS—TECHNIQUES AND TOOLS

The analysis and design of computer systems must therefore proceed in a disciplined manner with predetermined orderly steps using clearly defined techniques. The techniques are such that the output of one step is used as the input to the subsequent step or steps. A method can be considered to be an integrated collection of procedures, techniques, tools, and documentation aids organized

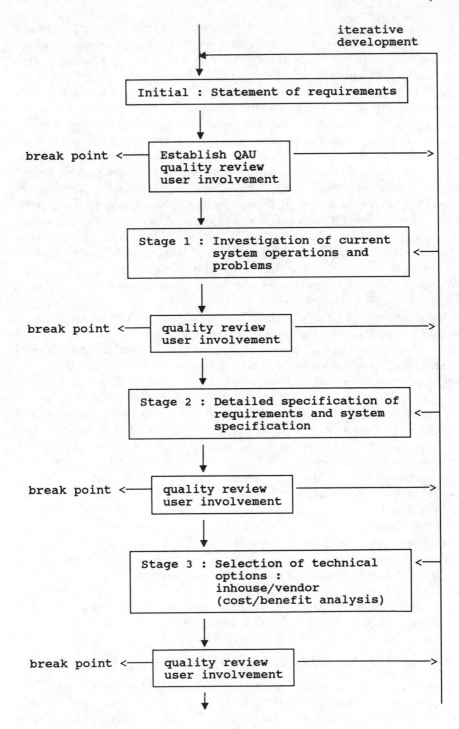

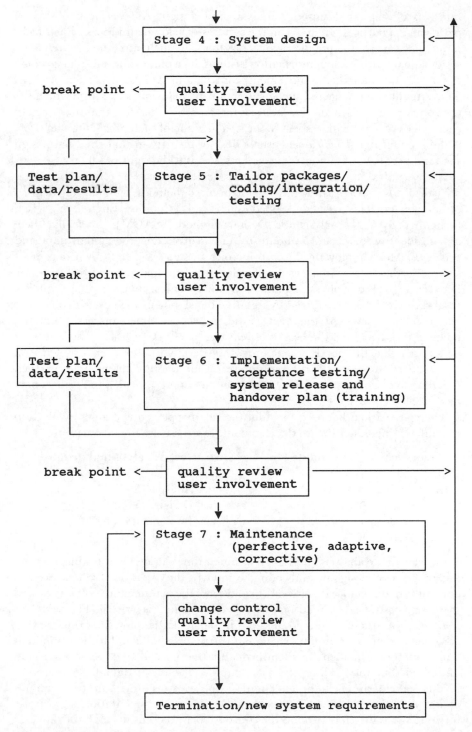

Figure 8.1 *System life cycle*

into sets of steps that support analysis and design. A *technique* is a way of performing a particular activity in the systems-development process. Each technique may involve the use of one or more tools. As such, a method represents a systematic way to develop information systems. The phases and sub-phases guide the user in the planning, management, control, and evaluation of such projects. The output of each stage is a specification for the following stage from which the appropriate design is made. *Verification* is the process of ensuring that the design of each stage is correct with respect to the specification of each preceding stage, *i.e.*, is the product right? *Validation* ensures design integrity in that the final design should satisfy the initial user requirements, *i.e.*, is it the right product?

In this context, a more complete system description can be obtained by what are called *structured methods*. Many such structured methods, with varying degrees of complexity, have been developed, such as Information Engineering (IE), Structured Systems Analysis and Design Method (SSADM). Typically, these aim to identify, map, and validate by cross-correlation the system data and processes. Most employ the basic principles of step-wise, top-down decomposition. Further, structured system development methods usually differentiate between logical and physical models. Quality control and, therefore, quality assurance are further enhanced by iterative development with user involvement and regular reviews. At the start of the project, a named project team that includes members from the *Quality Assurance Unit* (QAU) is established. It must be stressed that the development process is iterative, unforeseen problems (of which there are many!) will require reconsideration, refinement, and even redefinition of earlier stages. The *break points* are points at which the project may be cancelled (Figure 8.1).

There are various figures quoted for the relative percentage times involved in the different phases of system development, but these are my favourites:

Traditional analysis and design life cycle	Structured analysis and design life cycle
(i) Analysis, 5%	(i) Analysis, 20%
(ii) Design, 10%	(ii) Design, 35%
(iii) Implementation, 35%	(iii) Implementation, 35%
(iv) Testing, 50% plus!	(iv) Testing, 10%

With inadequate analysis and design as much time can be spent testing a system and programs as designing and coding. Similarly, the cost ratio of software development to maintenance is between 1:5 and 1:50; the maintenance costs far exceed the design costs. Code is read many more times than it is written (Figure 8.2).

Data driven structured design methods in particular employ the principle that all organizations have an underlying data structure that is stable. This inherent data structure can then be identified and the associated data access routes optimized. Suitable techniques, for example the use of data flow diagrams, further serve as an aid to documentation and assist as a medium of communication with the end users and external agencies. Documentation is typically intrinsic and is therefore produced as the designer proceeds through the different stages.

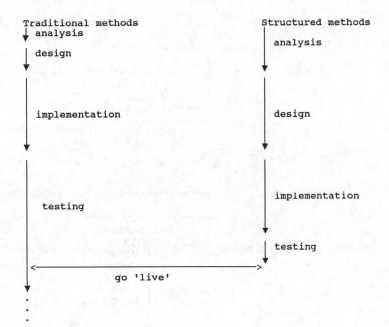

Figure 8.2 *Analysis and development time scales*

A variety of techniques common to many structured methods exist, for example, data flow diagrams (DFDs) and decision trees are used to describe processes. Entity modelling and normalization are techniques for describing and manipulating data. We will now briefly consider techniques for data and process description.

8.2.1 Process Descriptions—Data Flow Diagrams

Process description techniques are used to identify within the system all processes and data stores, and the associated data flows between them. Data sources and sinks external to the identified boundary are also noted. Data flow diagrams are made up of a variety of symbols used to represent the system components. Typically, there are four kinds of symbol used to represent processes, data stores, data flows, and external entities (Figure 8.3). Data flow diagrams are typically used for data, but it is not uncommon to apply this technique to material flow. In the initial stages of analysis it is often helpful to include material flows, the initial 'dirty' DFD being progressively refined in later stages.

Initially, the system is modelled by one process in a context diagram (Figure 8.4). This shows all the external entities that interact with the system under investigation and the associated data flows. In our case, the laboratory interacts with clients and the accounts department. Clients are sources of samples and the associated documentation. The laboratory is the source of accounting data that are supplied to the accounts department, which is a data sink. The context

Process

```
┌─────────────────────┐
│ 1                   │
├─────────────────────┤
│ Accept samples      │
│ and sample          │
│ certificates        │
└─────────────────────┘
```

Data flow

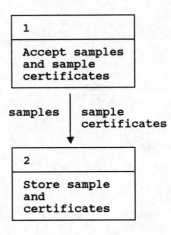

Data store

```
┌─────────────────────┐
│   rates schedule    │
└─────────────────────┘
```

External entity

```
( client )
```

Figure 8.3 *Data flow diagram symbols*

diagram does not describe the system in detail. Further detail is provided by examining the major system processes within the laboratory and drawing the initial DFD. The initial DFD for the current system identifies major system processes and the associated data flows contained within the system boundary (Figure 8.5). To aid cross-referencing, each process and document store is identified by a number.

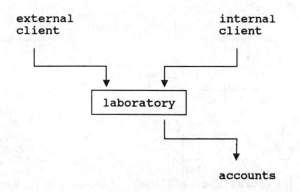

Figure 8.4 *Context diagram for a laboratory*

Processes show what systems do. Each process has one or more data inputs and produces one or more outputs. A file or data store is used to hold sample documents. Data can be retrieved from data stores. External entities are outside the scope of the area of interest but supply data into the system or use the output from the system. Data flows model the passage of data by using lines to connect processes.

By means of *diagram levelling*, all detail is controlled and incrementally introduced down to the process logic level. Levelling allows one to start at the top of the top-level function, the context diagram, and to add more detail progressively. These more detailed diagrams are called lower-level DFDs. Levelling improves the readability of the DFD. One important aspect of levelling is numbering. The context diagram is given the number 0, and processes in a top-level DFD are numbered consecutively, starting with 1. As each process is labelled, its DFD is given the same number as the parent process (Figure 8.6); the next more detailed DFD for process 3 would be 3.1. Another important aspect of levelling is data flow balancing—all data entering a process are the same as those entering its levelled DFD. As necessary, each process within the initial DFD can be expanded into a separate, more detailed DFD. Levelling continues until we reach the process logic level, where all processes in a DFD must have an associated process description. Process description techniques vary but include structured English, decision tables, decision trees, and action diagrams.

It is important to note that DFDs are not flowcharts, but are characterized by an absence of control elements, having no split data flows, having no crossing lines, conservation of data, and the use of meaningful names for processes and data flows. The use of these conventions and rules ensure that DFDs are self explanatory, complete, and unambiguous.

Structured system development methods typically differentiate between logical and physical models. The *logical model* defines what is done, and is implementation independent. This can be contrasted with the *physical model*, which maps how things are done. There are prescribed steps for converting a physical DFD to a logical DFD. In system analysis, a physical model is obtained

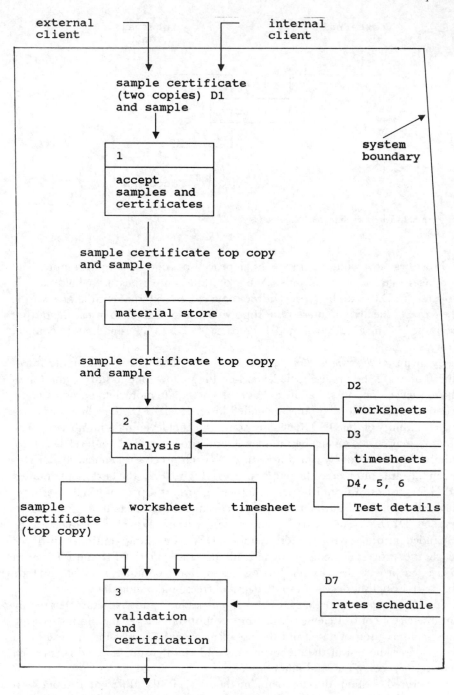

Figure 8.5 *Initial DFD for current laboratory system*

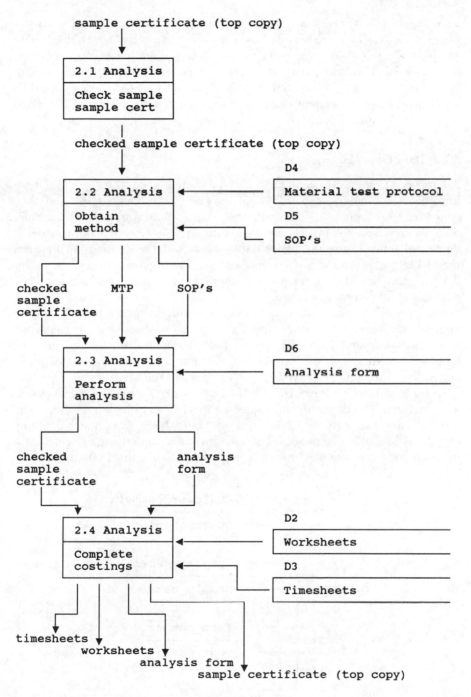

Figure 8.6 *Level 1 DFD for current system, Process 2: Analysis*

from the current system; the current logical model can be derived from this physical model. This analysis defines any problems and system objectives to complete the initial user requirements specification. In system design, a new logical model is produced that satisfies the previously defined system problems and objectives. The physical DFD is therefore converted to a logical DFD, implementation-independent model. The logical design model is then converted to a physical implementation of the new system.

8.2.2 Data Descriptions

An important aspect of systems analysis is the analysis of data. As with DFDs, an implementation-independent model of the data is developed first and then converted into a physical implementation. Data analysis is, however, a more difficult exercise than process analysis. Initially, a conceptual data model is developed representing the major data objects and the relationship between them. The normalization process that we have already met then organizes this data into a more optimal form. The last stage is the conversion of this normalized data model into a physical database.

One commonly used technique is entity modelling, which represents entities and the relationships between them. The pictorial representation is called the logical data struture (LDS). To draw a LDS, the following elements must be identified: entities, *i.e.*, distinct objects; relationships, *i.e.*, meaningful inter-actions; and attributes, *i.e.*, properties of entities and relationships.

In Figure 8.7, the entities CUSTOMER, SAMPLES, and TESTS are related. The next step is to define the attributes (properties) of objects in the sets. The attributes are then included in the diagram. For each entity a unique attribute, called the identifier, is chosen. The identifiers are underlined. Logical data structure diagrams are also used to express cardinality; this is the number of

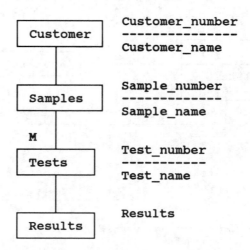

Figure 8.7 *Logical Data Structure (LDS) diagram*

relationships in which an entity can appear. For example, one sample may have a relationship with many tests. This is shown by the letter M on the diagram.

One output of the logical design is the LDS model of the new system. This model defines the data requirements of the new system and must be converted to a database implementation. This conversion is performed after the application of relational analysis to remove redundancies. To convert an LDS diagram into a set of relations, each set in the LDS diagram is replaced by a relational table. Database design is the process of converting the logical data design model to the appropriate structure supported by the DBMS, *i.e.*, database definition.

8.2.3 Documentation

Documentation consists of manuals, written procedures, policies, records or reports providing information concerning development, uses, maintenance or validation of a process or system, and any written or pictorial information describing, defining, specifying, reporting, or certifying activities, requirements, procedures, or results (ANSI N45.210–1973).

Documentation must be seen as both a tool of communication and of system maintenance. To be of most use, documentation must be mandatory with an agreed 'house' style. Documentation is a record of all the data gathered during analysis and design, and is organized in an 'easy-to-use' manner. Aids to ease of use include indexes, cross-referencing, and storage in one place. As such, documentation is often called a *system directory*. This directory is a repository for the data and process analysis diagrams.

8.3 VERIFICATION AND VALIDATION

Verification is the process of ensuring that the design of each stage is correct with respect to the specification of each preceding stage (is the product right?). Validation ensures design integrity in that the final design should satisfy the initial user requirements (is it the right product?). The importance of verification and validation can be demonstrated by the number of definitions that exist! Verification is:

'the process of determining whether or not the products of a given phase of the software development cycle fulfil the requirements established during the previous phase' (ANSI/IEEE Std 1012, 1986, and Std 729, 1983);

'formal proof of program correctness' (FIPS Pub 132, May 1988);

'the act of reviewing, inspecting, testing, checking, auditing, or otherwise establishing and documenting whether or not items, processes, services or documents conform to specified requirements' (ANSI/ASOC A3, 1979); and

'the demonstration of consistency, completeness, and correctness of the software at each stage and between each stage of the development life cycle' (FIPS Pub 1010, June 1983).

Validation is:

> 'The process of evaluating software at the end of the software development process to ensure compliance with software requirements' (ANSI/IEEE Std 1012, 1986, and Std 729, 1983); and

> the process of evaluating a system at the end of the development process to assure compliance with user requirements.

The difference between verification and validation is illustrated in Figure 8.8.

The aim of quality assurance is to design data systems that meet user specifications and regulatory standards. System verification activities are considered the responsibility of the QAU. Verification is required at all stages of all activities, from analysis and specification of requirements through design, programming, system integration, testing, implementation, and maintenance. Verification activities during, for example, the analysis phase include reviews and assessments of the specifications to ensure that the requirements are not only consistent, complete, and correct, especially in the context of company standards, but also comply with government regulations. The quality of the final system depends on the verification activities that took place during the system life cycle.

8.4 HOW TO DESIGN A GOOD PROGRAM

Even if a LIMS is used it is often necessary to produce code, for example, to interface to an instrument. The aims are to design programs that are easy to read, understand, and modify, all within acceptable timescales. To do this, it is necessary to improve the design and associated documentation of program logic, *i.e.*, structured or language-independent design. For this we can use *modular* programming. Modular programming is the means by which a problem/program is divided into separately named and addressable elements called modules that are integrated to satisfy the program requirements (Figure 8.9). This modular approach allows us to employ a stepwise, top-down decomposition of the problem. The stepwise refinement allows the deferment of detailed considerations by the use of abstraction to suppress and emphasise detail, as appropriate. More simply, you leave out the detail until it is required. Programs treated in this way are easy to understand and to upgrade, and provide no difficulties for the diagnosis of errors.

Traditional design methods are not suitable vehicles for high level language design. They very quickly become both complex and unmanageable. Using a good design method, for example, Jackson Structured Programming, it is possible to produce logically correct programs that are then converted to the target language. The coding process should only be the mechanical application of the rules of syntax of the target language. The code is only the means by which to

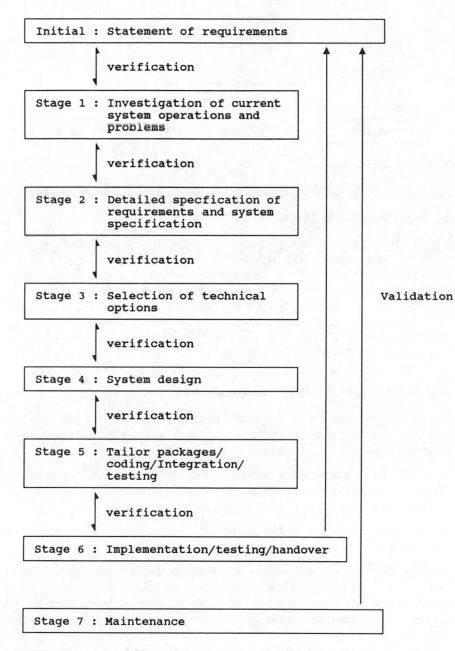

Figure 8.8 *Verification and validation*

communicate with the computer in order to execute the design. With incorrect or incomplete designs it becomes necessary to solve each problem as it arises, often generating other problems.

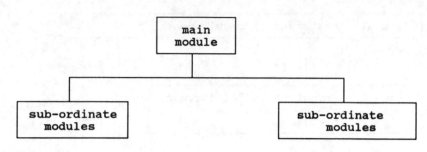

Figure 8.9 *Modular structure*

8.5 SOFTWARE RELIABILITY

The reliability of any system will depend on the following: the *completeness* of the analysis of the current system and its associated problems; the *correctness* of the system design; the correctness of the mapping between the system design and the implementation, sometimes called *interpretation*; and the *reliability* of the components of the system. We therefore depend on design and implementation correctness. Our program should meet its specification, never produce incorrect output, never allow itself to be corrupted, and take meaningful actions in unexpected situations. Certainly a tall order!

8.6 CRITERIA FOR GOOD PROGRAMS

What do we mean by a good program? Certainly authors of programs maintain that they write only good programs! Criteria for evaluation include readability, reliability, maintainability, robustness, and portability.

Readability. Programs should use meaningful variable names with controlled use of comments, and therefore be easy to understand. The code should be structured and modular.

Reliability. The program should be tested to demonstrate that it will serve the task it was designed for. Designing and testing are to a specification.

Maintainability. Is the code easy to maintain? For this to be the case one must easily be able to understand the current system, define clearly the required modifications, determine and minimize undesired effects of the modifications on the rest of the system, implement and document the modifications, and test the new system.

Types of maintenance include *perfective, adaptive,* and *corrective.* Corrective measures entail correcting any errors. Note that the correction of one error should not lead to others! Perfective changes provide new facilities that were not previously available. Adaptive modifications encompass environmental changes, *i.e.,* a change in the supporting hardware, operating systems or target language. It is worth noting that about 50% of the resources of a given department will be for maintenance.

Robustness. 'Press any key to continue'. What if the user presses the Break key? Does the program deal sensibly with unusual data?

Portability. Is the design specification portable between different machines, different operating systems, and different languages?

It is important to isolate the code from the effects of external modifications. In the final analysis it should be possible to move all the software from one computer system to another without rewriting it all.

Certainly other criteria for evaluating programs include factors that are concerned with performance such as CPU efficiency, memory efficiency, response time, *etc.*

8.7 MODULAR PROGRAMMING

Modularity is the extent to which software is composed of discrete components such that a change to one component has minimal impact on other components (ANSI/IEEE Std 729, 1983). When a modular solution is used, different levels of abstraction are considered. The highest level encompasses the solution in broad terms, *i.e.*, the problem environment, the lower level, procedural orientation, *i.e.*, the major software tasks, and the lowest level, direct implementation.

Each level is a refinement of the preceding one. We can therefore talk about *stepwise refinement* of a given problem. Stepwise refinement can be defined as 'a transformation applied to a single step of a process description in a sequence of one or more commands expressed in a form more closely related to the problem and the set of available executable primitives'. In this context it is important to achieve '*single mindedness*' in that each module should address a specific group of needs and have a simple interface when viewed from other parts of the program. It is then possible to have plug-in modules. A suitably designed module can be replaced by an equivalent module with little or no change to other parts of the program. This has obvious advantages for maintenance, portability, *etc.* This independence is measured by *cohesion* and *coupling*. Modules may then be converted to programming procedures. A procedure is a named portion of a computer program that performs a specific task. According to the ANSI/IEE Std 729, 1983, it is the course of action taken for the solution of a problem.

8.7.1 Cohesion

Cohesion is the measure of relative functional strength of a module or of how well the partitioning has been performed. With good cohesion, module replacement is simple—there are minimal side effects. Similarly, error detection and correction are easier in that the module can be identified quickly or eliminated as a source of the error. This can be contrasted with the other extreme of poor 'coincidental cohesion' in which the association between statements is minimal or zero. This would occur if a monolithic program were arbitrarily subdivided into modules.

8.7.2 Coupling

Coupling is a measure of the strength of the interconnection between or inter-dependence of modules. Low coupling indicates a well-partitioned system and is attained by the elimination of unnecessary relationships, thereby minimizing the effects of changing modules. Modules are uncoupled if they can function independently of each other. This is perhaps rarely the case. Modules have to interact and therefore must be coupled. There are different types of coupling, the strengths of which can be estimated.

The term 'minimal connection' is used to describe a fully parameterized transfer of control to a single entry point within a called module with an implicit return to a single point in the calling module. By 'fully parameterized' we mean that all information transfer between different modules is explicit or visible by module parameters. There are no hidden or implicit information flows via global variables. Global variables are available to all modules in the program. Are you sure no one is going to modify that global variable you are using? Having shared data makes error detection and correction much harder as potentially any part of the code can change any data value. Program modification is also made more difficult.

Coupling and cohesion are essentially linked. If there are multiple entry points this implies that the module has multiple functions and therefore low cohesion. Using single entry points implies a much weaker coupling between modules. The measure of the strength of coupling is the complexity of the interface between communicating modules. A module that has to import or export 10 parameters is more tightly coupled than one that involves the passing of one parameter. What is passed between modules has a bearing on the coupling strength. If data are passed between modules, there is a looser coupling than if control information passes between them.

8.7.3 Information Hiding

Information hiding suggests that modules can be characterized by design decisions that each module hides from all the others. A module is only aware of another if it has to, *i.e.*, the 'need to know' (Figure 8.10). Hiding implies that

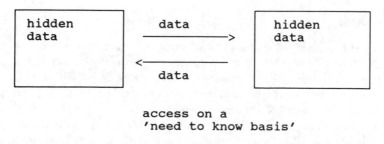

Figure 8.10 *Information hiding*

modularity can be achieved by defining a set of independent modules that communicate to one another only that information that is necessary to achieve the software function. Each module should be allowed access only to the variables that are
required to implement the functions appropriate to that module. Access to other
objects, if not needed, should be denied. Always remember that 'hidden information cannot be corrupted'. The resulting program is more secure with greater
data independence. Information is therefore a design criterion with implications
for testing, maintenance, *etc*. There is also better readability as the variables are
declared only where they are needed. For those new to programming, it is important to realize that programs are often developed by teams. Certainly, when you
are developing a program by yourself you know that one part of your program will
not corrupt the data used by another part!!! Consider therefore the situation where
there is a team of programmers.

8.7.4 Module Span

The consequence of stepwise refinement is the hierarchical structure that we
have met before. The span of control of a module is the number of subordinate
modules that it immediately calls (Figure 8.11). The suggested critical number is

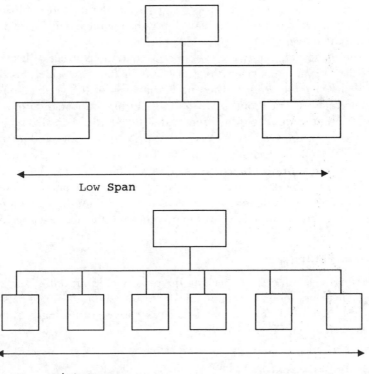

Figure 8.11 *Module span*

10 modules. A number in excess of this, *i.e.*, a high span of control, is the result of a failure to define intermediate levels correctly. This can be corrected by defining cohesive groups in order to form the intermediate level. A low span of control can be reduced by breaking the subordinate modules into smaller cohesive groups.

The aims for module design are to combine *high cohesion and low coupling with a medium span of control and optimum size.*

8.8 PROGRAM DEVELOPMENT AND TESTING

A major objective and outcome of structured design is to produce programs that are more likely to be correct. Design methods help us to produce 'correct' designs. The emphasis should be on the design process. The act of coding should then be a simple, mechanical process. Any errors at this stage will be trivial, *e.g.*, transcription errors. It is important to realize that 'bits do not rot', 'bytes do not decay', and the 'system is not stressed'. Software does not 'wear out'. Errors are introduced and errors propagate. The next time you purchase a software package there is every chance that it will be defective.

Errors can be introduced at all stages of development. Further, if errors are inherent in the design we could well be in the position of 'cementing-in' bugs. The aim of testing is to find errors. The algorithm for testing can be simplified to 'find errors early and find all errors'. It is essential therefore to incorporate testing as an integral part of the design, thus reducing the likelihood of designing errors into the system.

The verification of a software system is a continuous process through each stage of the software life cycle. Program testing is the most widely used system verification technique. Program testing is that part of the verification process that is normally carried out prior to and during implementation. Testing involves exercising the program using data similar to the real data that the program is designed to execute, observing the program outputs, and inferring the existence of program errors or inadequacies from anomalies in that output.

Do note that testing is the process of establishing the existence of errors. Debugging is the process of locating and correcting these errors. It has been said many times and bears repeating again, that 'testing cannot show that a program is error free'. Testing can only show the presence not the absence of errors.

8.8.1 Data Vetting

Data vetting is the process of checking the quality of input data. The program must be able to respond in a meaningful way to any data that can be presented to it. Data vetting procedures can therefore be extensive. Accordingly, data can be classified as shown in Figure 8.12.

8.8.2 Systematic Testing

The modular approach we have been using can be employed in the testing process. In this case it is possible to identify distinct stages: module testing; subsystem testing; system testing; and acceptance testing.

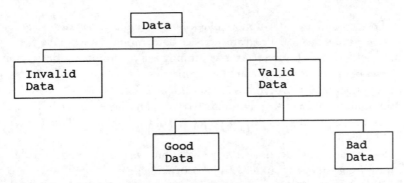

Invalid data : will cause systems corruption and therefore
               have unacceptable consequences.

Valid data   : can be dealt with by the program in an
               acceptable manner. Valid data can be
               either good or bad data.

Good data   : data that will be processed according to
               the program specification.

Bad data   : data that will not be processed but will
               be detected and the appropriate action
               taken.

Figure 8.12 *Data*

Module Testing. This is to ensure that a module performs to specification. It should be possible to test each module as a stand-alone entity.

Subsystem Testing. Modules are connected together and co-operate to perform a given series of tasks.

System Testing. Does the system provide the functions specified in the acceptance document?

Acceptance Testing. Prior to acceptance testing all testing and test data are artificial. Acceptance testing is the process of testing using real or live data.

There are two main methods for testing programs, *black box* testing (also referred to as *stress testing*) and *white box* testing (also known as *glass box*). They are complimentary rather than exclusive testing methods.

8.8.3 Black Box Testing

In this approach the program is seen as a 'black box'. It is a piece of software the contents of which are unknown to the tester. However, the specification is known and the functions and external interfaces can therefore be tested. Black box testing attempts to detect errors in the following categories: initialization and termination errors; incorrect functions; missing functions; module interface errors; and data structures. Input data are generated systematically and the associated output results are inspected accordingly.

The basic ideas that support black box testing are *equivalence partitioning* and *boundary value analysis*. Programs expect data input. As we have seen, this data may be classified. Equivalence partitioning is a technique for determining which classes of input data have common properties. It is then possible to test only one member of that set. Our aim is to establish a reasonable level of confidence with a minimum set of test cases. For a given data item there will be a defined range of values. An equivalence class is a collection of values that will have the same effect on the program.

System test cases can include valid input, *i.e.*, good and bad data, invalid input, *i.e.*, system corruption, and performance.

Consider the simple case where a program expects data in the range 0 to 4 inclusive. There are four equivalence classes: (i) input data < 0, bad data; (ii) input data $\geq$ 0 and $\leq$ 4, valid, good data; (iii) input data > 4, bad data; and (iv) input data are non-integer values, bad data.

Using equivalence partitioning we can identify that $-1, -2, -3$, *etc.*, belong to the set of data corresponding to the first class. Similarly, 0, 1, 2, 3, and 4 belong to the same (the second) class. In order to test our program, at least one member from each set should be included. This may seem trivial but what if the input data are 0 to 1000?

It is often the case that errors are due to equivalent class boundary values, typically, a misunderstanding regarding the use of 'greater than' rather than 'greater than or equal to'. In boundary-value testing, the test data consist of the boundary values. In our example, valid data are 0 to 4 inclusive. The boundary values are therefore -1 (bad data); 0 (valid, good datum); 4 (valid, good datum); and 5 (bad data).

Other data validations include the following: incorrrect field length in which the input data do not match the required length of data field—either the input data fields are too long or too short; alphanumeric validity where alphabetic instead of numeric characters are input, or *vice versa*; and inconsistent data, sometimes called illogical data, in which the data are within the range but incorrect.

8.8.4 White Box Testing

In this approach we know the internal workings of the program. The control structures of the program are used to design test cases. Typically, we must exercise all logical selection decisions. TRUE and FALSE selections mut be made for all unary and binary control constructs. Similarly, all CASE options must be exercised. All iterations must be exercised, both at their boundaries and within normal operations, and internal data structures must be exercised.

Knowing the specification, we can conclude that testing to see if the program will meet the specification is all that is needed. Can all the sequence, selection, and iteration routes be tested?

Consider for a moment an automatic washing machine program:

10 'programs'	$10 \times$
Biological or non-biological	$2 \times$
Full- or economy-load	$2 \times$
Extra spin option	$2 \times$
No rinse option	$2 \times$

This gives rise to 160 combinations. Consider now a multi-user operating system with some users multitasking. The total number of combinations is more or less incalculable.

8.9 PROJECT FEASIBILITY—WHAT WILL IT COST?

A project-feasibility study determines the return from investment in a project. To perform such a study, it is necessary to determine the cost of purchases and activities needed to implement the project. Furthermore, value must be assigned to the benefits that may be gained from the use of the new system. The economic feasibility of a project can then be determined from a cost/benefit analysis. There are two main steps involved in such an analysis, estimation of the costs and benefits, and estimation of the value of the project given the cost/benefit analysis.

The cost/benefit analysis is typically performed for a 5 year system-life period. A shorter period would not allow the new system to become fully effective. For a longer period, the estimates and projections are unpredictable.

8.9.1 Estimating Costs

Cost estimation is not an exact science. Some items are tangible and as such can be directly valued, for example, the following: equipment costs (purchase, lease or rental)—computers, peripherals, modems, air conditioning, UPSs, security systems; installation costs—structural alterations, ducting routes; development costs—consultation fees, software development; personnel costs—staff recruitment and relocation, training, redundancy payments; operating costs—consumable materials, rent, rates, depreciation, maintenance, insurance; equipment costs for the new system—computer, printers, *etc.*; personnel costs—staff required to develop and maintain the new system; and running costs—stationery, documentation, *etc.*

The cost of intangible items, however, cannot be easily or precisely determined as there is considerable subjectivity involved. For example, what is the cost of the laboratory managers' time when assisting in the implementation of a new computer system? Each of the above costs is estimated on an annual basis for the first 5 years. This estimate can be repeated for each project proposal.

8.9.2 Estimating Savings

The benefits can also be classified as tangible or intangible. Tangible benefits include the following: personnel savings—reduction in the number of staff due to

Table 8.1 *Comparison of two systems in terms of annual savings*

| | Annual cost/$1000 | | |
Year	Current system	Proposed system	Annual savings/$1000
1	50	100	− 50
2	50	30	20
3	50	30	20
4	50	30	20
5	50	30	20

automation; and operating savings—greater sample throughput. Intangible benefits relate, for example, to the following: planning and control information—readily available statistical information, better sales due to the improved response time in answering queries and better job scheduling; and personnel—staff are freed from routine tasks.

The estimated savings are calculated on an annual basis for the first 5 years. The costs and benefits are then used to evaluate the economic viability of the proposed system. It is essential that the new system saves money, either by reducing costs or increasing profit. There are various investment appraisal methods for determining the economic viability of a proposed project. Assume that the existing system has an annual cost of $50 000 per annum and that the proposal is to replace it with a new system. The new system requires an initial outlay of $100 000 and thereafter has an annual cost of $30 000 (Table 8.1). The choice of system depends on two factors, the cost of capital and the system-life period. The cost of capital is the current lending rate.

8.9.3 Payback Method

The payback method is a simple, but crude calculation used to determine the time taken to recover the initial investment. The method takes no account of the savings at the end of the payback period. Using the example from Table 8.1, the payback period is $50 000/$20 000 = 2.5 years.

8.9.4 Returns to Outlay Ratio

This simple method takes into account the accrued savings over the system life. In the previous example, the returns to outlay ratio are (4 × $20 000)/$50 000 = 1.6. However, this is still a crude calculation, as no consideration is given to the timings of the savings.

For more realistic calculations, the discounted cash flow (DCF) must be taken into account. When profit is the sole consideration, the proposed project must attract a return at least equivalent to the current investment rate. The main concept is that a given sum of money received today is worth more than the same amount received at sometime in the future. This is because today's money can be

Table 8.2 *Calculation of the cumulative net present value*

| | Annual cost/$1000 | | | | |
Year	Current system	Proposed system	Annual savings*/$1000	Discount factor†	Present value‡/$1000
1	50	100	− 50	1.0	− 50
2	50	30	20	0.870	17.4
3	50	30	20	0.756	15.12
4	50	30	20	0.658	13.16
5	50	30	20	0.572	11.44

* Net cash flow.
† 15%.
‡ Cumulative net present value = − 50 + 17.4 + 15.12 + 13.16 + 11.44 = 7.12.

used to earn money in that time. All future money must therefore be discounted by a given discount factor, the value of which depends on the country's economy. As a simple example, $1000 due in 1 year at a discount rate of 15% is worth $1000/1.15 = 870 today. Discount factors for different rates and periods, are available from discount tables, however, they can be simply calculated from the repeated division of 1 by $(1 + r)$, where r is the discount rate. To a first approximation

$$NPV = ncf + ncf/1.15 + ncf/1.15^2 + ncf/1.15^3 \ldots$$

where NPV is the net present value, and ncf is the net cash flow.

Though laboratory automation is often seen as a service rather than a profit-making exercise, DCF is used in assessing the cost to benefit ratio. One method that uses DCF is the net present value.

8.9.5 Net Present Value

The NPV method discounts all future cash flows (expenditure and savings) back to today's value. The sum of these values is called the cumulative net present value of the system—the higher this figure, the better (Table 8.2).

Note that no account has been taken of factors such as capital allowances, corporation tax, *etc.* A tabulation of the complete calculation is easily done and is invaluable in presenting a case for laboratory expenditure (Table 8.3).

Table 8.3 *Skeleton table for the calculation of cumulative net present value*

	Initial	Year 1	Year 2	. . . . .
Costs				. . . . .
				. . . . .
Equipment				. . . . .
Installation				. . . . .
Development				. . . . .
Personnel				. . . . .
Operating				. . . . .
Intangible				. . . . .
Total costs				. . . . .
Savings				. . . . .
				. . . . .
Equipment				. . . . .
Personnel				. . . . .
Operational				. . . . .
Allowances				. . . . .
Intangible				. . . . .
Total savings				. . . . .
Net cash flow (savings − costs)				. . . . .
Discount factors				. . . . .
NPV				. . . . .
Cumulative NPV				. . . . .

CHAPTER 9

Analysis, Design, Implementation, and Testing of Laboratory Information Management Based Systems

9.1 GOOD LABORATORY PRACTICE

During the period from the 1960s to the 1970s the US regulatory agencies were faced with a number of discrepancies in laboratory data that had been submitted to them. There were instances of laboratories not following protocols, and a lack of documented Standard Operating Procedures, SOPs; if SOPs were available, they were poorly complied with. It was noted that several laboratories had the general problem of having poor documentation and making incomplete data reports to regulatory agencies. The response of the US Food and Drug Administration (FDA) was to propose the Good Laboratory Practice (GLP) regulation in 1978. The Environmental Protection Agency (EPA) also joined this initiative. In recognition of the importance of the international chemical trade, the FDA and EPA joined with other countries to address these issues on an international basis. The requirement for standardization within the member countries of the Organization of Economic Cooperation and Development (OECD) led to the formation of an OECD Expert Group in 1978. The remit of this group was to develop international guide-lines for good laboratory practice and to assure that the data developed in one country would be acceptable to all other member countries. In 1981, there was a formal decision by OECD members to accept, within practical legal limits, data generated according to the 'OECD Test Guide-lines' and the 'OECD Principles of Good Laboratory Practice'. Two additional documents were produced: 'Implementation of OECD Principles of GLP'; and 'Guide-lines for National GLP Inspections and Study Audits'. These documents encouraged OECD members to adopt GLP as part of their legislative procedures and to implement laboratory inspections in order to monitor compliance with the principles of GLP.

In the first instance, the regulators were unable to envisage the speed at which computers would become an essential part of laboratory procedures. Computers are now standard items of laboratory equipment. Ultimately, the quality of computer systems depends on the skill, training, and ability of the people who design, install, operate, and maintain them. In the pharmaceutical industry there is little margin for error and hence this is a highly regulated manufacturing industry. As such, a series of compliance guides have been published by the FDA:

'Compliance Policy Guide #7132a.07', FDA, October 1982, Chapter 32a: Computerised Drug Processing: Input/Output Checking;

'Compliance Policy Guide #7132a.08', FDA, December 1982, Chapter 32a: Computerised Drug Processing: Identification of Persons on Batch and Control Records;

'Guide to Inspection of Computerised Systems in Drug Processing', FDA, February 1983 ('Blue Book');

'Compliance Policy Guide #7132a.11', FDA, December 1984, Chapter 32a: Computerised Drug Processing: CGMP Applicability to Hardware and Software;

'Compliance Policy Guide #7132a.12', FDA, January 1985, Chapter 32a: Computerised Drug Processing: Vendor responsibility;

'Compliance Policy Guide #7132a.15', FDA, April 1987, Chapter 32a: Computerised Drug Processing: Source Code for Process Control Application Programs; and

'Software Development Activities. Reference Materials and Training Aids for Investigators—Technical Report', FDA, July 1987.

Following publication of the FDA's 'Blue Book' in 1983, the Pharmaceutical Manufacturers Association (PMA) established a Computer Systems Validation Committee (CSVC) to provide working guide-lines for validation of computer-based systems and to address controversial issues. The Pharmaceutical Manufacturers Association is a trade association representing most of the drug manufacturers in the US. As a result the CSVC published 'Validation Concepts for Computer Systems Used in the Manufacture of Drug Products' in May 1986. In an attempt to address the problems of computer terminology this document included a lexicon. The CSCV also elected to describe computer-system validation in the form of a system development life cycle (SDLC) (Figure 9.1).

The CSCV defines a computer system as consisting of six subsets (Figure 9.2); however, it fails to mention the role of people. As we will see, modern design methods address this failing.

After initial success with the original CSCV publication it became apparent that software and documentation needed greater attention; this resulted in the publication, by the FDA, of 'Software Development Activities. Reference Materials and Training Aids for Investigators—Technical Report', July 1987. Since then, the CSCV has continued to publish material concerned with software development and testing.

Irrespective of whether software is purchased or written in-house, the end-user plays a major role in the development of the requirements specification and the final acceptance testing. It is considered that a strong multidisciplinary team is fundamental to success. Document verification and over-all system validation provide quality assurance.

Both the FDA and the PMA define that user firms should possess and be

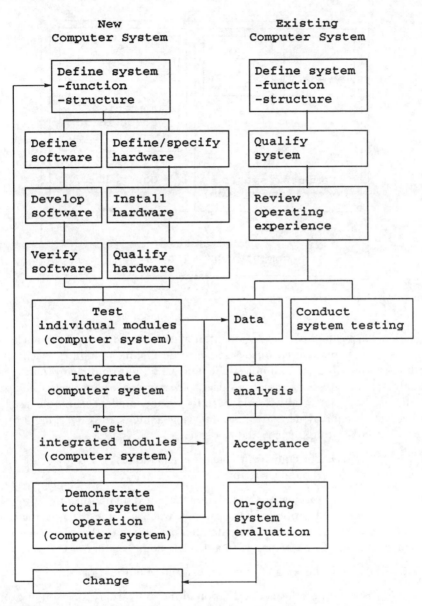

Figure 9.1 *System development life cycle*

familiar with application specific source code. This does not extend to operating system coding. However, both organizations initially failed to address the need to validate configurable software, *i.e.*, LIMSs, spreadsheets, *etc.* This issue is now being addressed through agreements with the vendors of such software.

The importance of testing is stressed throughout all stages of software development with a software testing policy stating the organization's standards. The standards must state how compliance is to be measured, and identify responsible

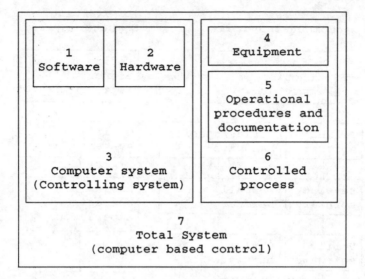

Figure 9.2 *Computer system sub-sets*

individuals/departments, the documentation to be used, and the tasks to be performed. Test plans used during software development outline specific test methods, responsibilities, and objectives. The global test plan co-ordinates individual test plans. As we have seen, a variety of test methods have been developed. In all cases, all test results should be fully documented. Techniques used in testing include equivalence partitions, data flow analysis, desk checking, inspection, white box testing, review, and test case.

Equivalence Partition. The equivalence partition technique is also known as boundary value analysis or boundary testing.

Data-flow Analysis. Data-flow analysis verifies that the data handled by a program are defined and used properly. For example, when a value is passed between software modules, an error may be generated if a different variable type is used involving a change from an integer to a floating point number or *vice versa.*

Desk Checking. Desk checking is a manual inspection of the program.

Inspection. Inspection is a formal evaluation of the software by a person or group other than the author of the software.

White Box Testing. White box testing is also known as path analysis.

Review. A review is a formal meeting to evaluate the software.

Test Case. A test case is a specific set of test data and associated procedures developed for a particular objective such as to verify compliance with a specific requirement.

Testing strategies employ a variety of testing techniques. Different test strategies are used throughout the SDLC.

Acceptance Testing. Acceptance testing is a formal test to determine whether or not a system satisfies its acceptance criteria and to enable the customer to determine whether or not to accept the system.

Performance Testing. Performance testing verifies that the system has the appropriate data throughput, within defined limits, to handle normal and peak-load demands. The response time is often included in performance testing.

Stress Testing. Stress testing involves operating the system at its operational limits for extended periods.

Security Testing. Security Testing verifies that the appropriate security measures have been installed and are operational.

Recovery Testing. Recovery testing confirms that, should there be a system failure, it would be possible to return to a fully operational state within an acceptable period of time.

Installation qualification (IQ) of computer-based systems represents a systematic approach to the installation and checking of system hardware, which includes all computer-based equipment, computer peripherals, instruments, wiring, *etc.* It is essential to consider the complete system including field instruments. It should be noted that the trend is towards process in-line instrumentation where rugged analytical instruments are used in the production environment. As such, field instruments are subjected to thermal and mechanical shock, corrosive vapours, and electrical noise. The National Electrical Manufacturers Association (NEMA) classifies 18 different types of enclosure that provide mechanical and environmental protection. Incorrectly installed signal wiring is a major source of data errors. In particular, long cable runs using low-voltage electrical lines are vulnerable and as such should be shielded from EMI sources such as power lines, motors, *etc.*

Change control is the systematic approach to implementing changes in a system to ensure that a validated system remains validated.

Another publication of note is 'Computerised Data Systems for Non-clinical Safety Assessment—Current Concepts and Quality Assurance', which was first published in 1988 as the result of a meeting of selected experts for industry, government, and academia. It was a regulatory initiative sponsored jointly by the Office of Regulatory Affairs (US), the FDA, and the National Centre for Toxicological Research (NCTR). It is not mandatory to use the more detailed guide-lines in this document, but it does provide useful complimentary reading. It is sometimes referred to as the 'Red Apple Book'.

The following is a key document in the UK: 'Good Laboratory Practice, the United Kingdom compliance program', Department of Health, 1989. This states that: 'GLP is concerned with the organisational processes and the conditions under which laboratory studies are planned, performed, monitored, recorded, and reported. Adherence, by laboratories, to the Principles of GLP ensures the proper planning of studies and the provision of adequate means to carry them out. It facilitates the proper conduct of studies, promotes their full and accurate reporting, and provides a means whereby the integrity of the studies can be verified. The application of GLP to studies assures the quality and the integrity of the data generated and allows its use by Government Regulatory Authorities in hazard and risk assessments of chemicals'. This document is in accordance with international practices published by the OECD in the document 'Good laboratory practice in the testing of chemicals—final report of the OECD Expert group on Good Laboratory Practice', ISBN 9 264 12367 9.

In recognition of the increasing and widespread use of computers in laboratories for data capture and processing, a complimentary document exists: 'Good laboratory practice—the application of GLP principles to computer systems (advisory leaflet number 1); Department of Health, 1989.

Together these publications are sometimes referred to as the 'Blue Books'. The 1989 'GLP Monitoring Programme' of the UK's Department of Health expects conformance with these documents.

In the context of harmonization, the European Community (EC) produced directives that place obligations on member states to comply accordingly. For example, the directives 65/65/EEC, 75/318/EEC, and 75/319/EEC prescribe the requirements for controls over the manufacturing and testing of medicinal products. The directive 89/341/EEC, which amends directive 75/319/EEC, states that '. . . the quality of the medicinal products manufactured within the community . . . should be maintained by compliance with the principles of Good Manufacturing Practice (GMP)'. The EC GMP directives 91/412/EEC and 91/356/EEC are concerned specifically with veterinary and human medicinal products. There are five volumes for the rules governing medicinal products. Volume IV, in particular, is concerned with GMP: 'The Rules Governing Medicinal Products in the European Community', 1992, Vol IV: 'Good Manufacturing Practice for Medicinal Products'.

There are 12 annexes to the EC 'Guide to GMP' of which number 11 is concerned with computerized systems. The main issues are as follows: definitions; principles; personnel; validation and the life cycle; equipment siting; documentation; quality assurance of software development; testing and validation; controls (data confidentiality, integrity, availability); change controls (revalidation); hard copy; and qualified person.

In effect, GLP is concerned with quality and the improvement of quality. Quality is 'the totality of features and characteristics of a product or service that bears on its ability to satisfy given needs' (ANSI/ASOC A3, 1979). Quality assurance is the 'planned and systematic pattern of all actions necessary to provide adequate confidence that the item or product conforms to established technical requirements' (ANSI/IEEE Std 730, 1981). The major principles of GLP are listed.

Adequate Facility. The facility should be adequate to the extent that the work can be performed properly and safely.

Personnel. Personnel should be suitably qualified and trained.

Maintained and Calibrated Equipment. Records of calibration should be maintained, either separately or as part of the analytical sequence.

Standard Operating Procedures. Standard operating procedures are written procedures that can be followed by any well-informed, qualified person in the complete expectation that the anticipated result will be obtained.

Quality Assurance. Quality assurance concerns checks by independent qualified personnel. The QAU has the responsibility of double checking the procedures and results in order to provide assurance that the work is being properly conducted and can therefore be relied upon. The QAU uses SOPs to check the procedures and is obliged to countersign that the work is in compliance with GLP regulations.

In order to address these issues, QAU or 'quality assurance function' must be established in each laboratory. They also specify that certain tasks should be carried out by this unit or function. These requirements have resulted in the creation of a new scientific, managerial professional, the QAU manager.

9.1.1 Special Considerations for the Quality Assurance Unit

The QAU is directly responsible for assuring the management of quality. Hence, it is concerned with the facilities, equipment, personnel, methods, practices, records, *etc.*, that are governed by GLP regulations. The QAU has mandated responsibilities under the GLP regulations for SOPs, training, completeness of data, security, raw data, and calibration.

9.1.1.1 Standard Operating Procedures. The GLP regulations require the preparation of written SOPs that assure the quality and integrity of generated data. In the case of computer-based systems, the SOPs should address policy and practice for system development and performance, documentation, and document retention. The SOPs should describe the methods and schedules for routine inspection and maintenance of computer hardware and computer associated environmental equipment. They are also required for the handling, storage, security, back-up, archiving, maintenance, retention, and retrieval of computer-generated data. Security issues, both logical and physical, system controls, access controls, change controls, and disaster-recovery plans should also be addressed. The SOPs must include the details of training programmes, and the training requirements of all personnel involved in the design, production, and use of computer systems. The SOPs of the QAU must describe the methods and schedules for the QAU audit, and for inspection activities associated with both the development and use of computer-based systems.

9.1.1.2 Training. The introduction of computer-based systems into regulated environments necessitates the training of all personnel involved in the generation and use of these systems.

9.1.1.3 Completeness of Data. It is the responsibility of the QAU to examine the audit trail and any associated records for completeness and accuracy. In the case of system failure, the manual procedures for re-entering data should ensure that the data are complete and correct.

9.1.1.4 Security. This is concerned with the over-all protection of hardware and data against unauthorized or accidental modification, destruction or disclosure.

9.1.1.5 Raw Data. The GLP regulations define raw data as 'any laboratory worksheets, records, memoranda, notes or exact copies thereof, that are the result of original observations and activities and are necessary for the reconstruction and evaluation of the report of that study'. The GLP requirements for data entered into and stored in a computer are no different from those for hand-recorded raw data. Regardless of the mode of storage, the QAU must be able to review the data to allow assessment.

9.1.1.6 Calibration. Calibration is concerned with laboratory procedures and equipment calibration. There must be procedures concerned with equipment-calibration management, *i.e.*, recalibration dates, *etc.* The GLP regulations define the SOP and the equipment log-book. Each instrument may be assigned a log-book to record authorized instrument management. The SOPs are the agreed procedures as to who does what, and when. You are encouraged to read the source documents, as appropriate, but it is worth noting some practical implications of the GLP guide-lines: only authorized individuals can make data entries; data entries cannot be deleted; data changes must be in the form of amendments; the LIMS must be as 'tamper proof' as possible; SOPs must describe procedures for ensuring data validity; the individual responsible for direct data entry input shall be identified at the time of data input; any changes shall be made so as not to obscure the original entry; the change shall be dated; the person responsible for the change shall be identified; all raw data, documentation, protocols, and specimen and final reports generated as a result of non-clinical laboratory study shall be retained; there shall be archives for orderly storage and expedient retrieval of all raw data, documentation protocols, specimens, and interim and final reports; and all the above points shall be built into the SOPs.

For regulatory purposes, computer systems must comply therefore with GLP principles. The goal is to produce automated data systems that meet the user requirements and maintain data integrity. In order to achieve these goals, specific verification and validation activities are stressed by the GLP regulations. As such, the operations that involve the use of computers must comply with GLP principles and will be inspected to ensure that they are suitable for the purpose for which they are intended, procedures exist for adequate control and maintenance, and they are operated in a manner compliant with GLP principles. The conclusion is that there must be assurance that: 'suitable procedures exist to ensure that computer systems are suitably designed, controlled, operated, and maintained in order to properly accommodate the functions and activities to which they are dedicated.'

In 1988, Logica, UK published a Department of Trade and Industry (DTI) sponsored report on Quality Management Systems (QMS) for software. It was concluded that ISO 9001 was the best available standard; however, reservations were expressed. This was followed in 1989 by the British Computer Society reporting to the DTI a software sector certification scheme leading to the assessment and certification of software QMSs to ISO 9001. The final result was the 'Guide to Software Quality Management System Construction and Certification using EN29001', which was published in 1990, and is concerned with the construction and assessment of software QMS to ISO 9001. Updated in 1992, it has five main parts: part 1, Introduction, provides an introduction to QMS and certification; part 2, ISO 9000–3 Guide-lines for the Application of ISO 9001 to the Development, Supply and Maintenance of Software, outlines the requirements of ISO 9001 for software systems; part 3, Purchasers guide, indicates purchasers' expectations of assessed and certified software, and is in effect a summary of the Software Tools for Application to large Real Time Systems (STARTS) manual, which will be referred to later; part 4, Suppliers guide, provides guidance to both suppliers and in-house developers of software to ISO 9001 standard; and part 5, Auditors guide, provides guidance for auditors for conducting assessments of suppliers seeking certification of their QMS to ISO 9001.

9.2 GOOD MANUFACTURING PRACTICE AND GOOD LABORATORY PRACTICE

Computers, like any other equipment used in manufacturing, must function in a specified manner. The FDA GMP regulations (part 211.68) require that computer-based systems are routinely calibrated and checked accordingly to ensure correct performance. This has to be demonstrated explicitly (Figures 9.3 and 9.4). Part of the GMP regulations is concerned with laboratories and laboratory computers used in relation to manufacturing. In particular, part 211.68, Automatic, Mechanical and Electronic Equipment, states that such equipment 'shall be routinely calibrated, inspected, or checked according to a written program designed to assure proper performance'. Accordingly, such test records must be maintained with provisions for authorized change approval. Furthermore, all I/O must be verified with appropriate back-up procedures, all in the context of adequate security measures. More specifically, part 211.160, Laboratory Controls, General Requirements, relates to the general requirements for all laboratory systems, including computer systems, and states that 'Laboratory controls shall include the establishment of scientifically sound and appropriate specifications, standards, sampling plans, and test procedures designed to assure that components, drug product containers, closures, in process materials,

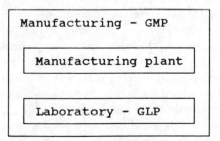

Figure 9.3 *GMP and GLP*

```
┌─────────────────────────────────────────────────────────────┐
│  Laboratory - GLP                                             │
│  ┌──────────────────────────────────────────────────────┐   │
│  │ Hardware                                               │   │
│  │ --------                                               │   │
│  │ computers : processors, memory hierarchy,              │   │
│  │             interface boxes, A/D                       │   │
│  │                                                        │   │
│  │ instrumentation : analytical instrumentation           │   │
│  │                                                        │   │
│  │ equipment : Modems, wiring, power supplies,            │   │
│  │             TV monitors, fire proof cabinets,          │   │
│  │             environmental controllers, etc             │   │
│  └──────────────────────────────────────────────────────┘   │
│  ┌──────────────────────────────────────────────────────┐   │
│  │ Software                                               │   │
│  │ --------                                               │   │
│  │ Operating system                                       │   │
│  │ Application packages (supplied, in-house)              │   │
│  │      E.g. databases, spreadsheets, etc                 │   │
│  └──────────────────────────────────────────────────────┘   │
│  ┌──────────────────────────────────────────────────────┐   │
│  │ Operational control                                    │   │
│  │ --------------------                                   │   │
│  │ Specifications                                         │   │
│  │ Standard Operating Procedures and system               │   │
│  │ controls                                               │   │
│  │     - security                                         │   │
│  │     - audit trails                                     │   │
│  │     - recovery procedures                              │   │
│  │ Data acquisition, processing and storage               │   │
│  │ Report generation                                      │   │
│  │ Documentation                                          │   │
│  │ Training                                               │   │
│  └──────────────────────────────────────────────────────┘   │
└─────────────────────────────────────────────────────────────┘
```

Figure 9.4 *GLP*

labelling, and drug products conform to the appropriate standards of identity, strength, quality and purity. Laboratory controls shall include: calibration of instruments, apparatus, gauges and recording devices at suitable intervals in accordance with a written program containing: specific directions, schedules, limits for accuracy, limits for precision and provision for remedial action'. Part 211.165, Testing and Release for Distribution, states, 'The accuracy, sensitivity, specificity, and reproducibility of test methods employed by the firm shall be established and documented'. Such validation and documentation may be fulfilled by part 211.194, which states that 'the suitability of all testing methods shall be verified under actual conditions of use.'

The FDA definition of validation is: 'establishing documented evidence which provides a high degree of assurance that a specific process will consistently produce a product meeting its predetermined specifications and quality attributes'.

The FDA applies this to all aspects of the manufacturing process. In this context, hardware validation is the demonstration of 'a high degree of confidence that the system will consistently do what it is supposed to do'. Similarly, software validation provides 'assurance that computer programs will consistently perform as they are supposed to within pre-established operational limits'. In this context, the inspectors pay particular attention to software and systems development.

9.3 PROBLEMS AND FAILURE

I was not sure what to call this section. What will be stressed is that problems will arise and, though for obvious reasons this is not made public, that new systems do sometimes fail to meet expectations for a variety of reasons—inflexible software, incorrect selection of hardware, poor response times, *etc.* This is in keeping with the findings of the National Computing Centre, which tolerate repeating: 'Inadequate or incomplete analysis results in systems that do not satisfy user requirements. Poor design produces systems that are inefficient, ineffective, and inflexible. The consequences are increased cost, delayed implementation, error prone operation and open-ended maintenance'. Also, the time taken to have a fully implemented automated laboratory should not be underestimated. A rule that invariably holds true is, estimate how long it will take, then double it, then double it again! If you are unlucky you will have to double again!! Regulations address the broad areas of responsibility, but provide little or no guidance on their direct implementation. For example, the guide to Good Pharmaceutical Practice, HM Stationery Office, London, 1983 states the following: '16.2 A general description of the (computer based) system should be produced and kept up-to-date. It should describe the principles and main features of the way in which the computer is used and how it interacts with other systems and procedures . . . 16.4 Systems should be designed to take into account the need for checks, at appropriate intervals, for correct operation'.

In regulated environments, the regulations provide guidance in the form of major principles and highlight the need for the QAU, which has the responsibility of checking the procedures and results in order to provide the assurance that the work is being properly conducted and can be relied upon. Unfortunately, the common experience is to be given responsibility without guidance other than the regulations, which, as we have seen, provide little or no guidance on their direct implementation. Laboratory Information Management Systems are essential components of an automated laboratory and provide many of the required features such as access rights, data input validation, *etc.* The expectation is that LIMSs will solve all problems. However, LIMSs in themselves do not provide a complete solution to laboratory automation problems.

Often there will be little or no analysis performed prior to the installation of the LIMS. Neither will there be a systematic approach to the analysis and design of the new system including data flows and new working procedures. Though the regulations define the need for a QAU, due to obvious limitations in cost and manpower, in small installations this is a dual role of either the quality or the laboratory manager. Considerable difficulties can be experienced with regard to the application of the GLP principles to the development of computer-based systems in these laboratories. Problems include: system development and implementation with inadequate verification and validation; insufficient documentation and inconsistent documentation standards; inadequate procedures and the absence of schedules for the routine inspection and maintenance of computer hardware and computer associated environmental equipment; inadequate procedures and schedules for data back-ups; inadequate recovery procedures; inadequate security controls; inadequate training programmes for analytical personnel involved in the design, production and use of the computer systems; incomplete SOPs for the operation of the computer system; and no consideration given to possible future integration with other on-site databases.

Inadequate attention given to the analysis and design of laboratories prior to the introduction of a LIMS can result in a computer system that does not fully satisfy user requirements—things can go wrong! Irrespective of the type of environment, it is essential to be able to obtain an accurate understanding of the information system and the associated dynamics of its operation prior to introduction of the LIMS. System dynamics include the data and how these data are used by the system and subsystem components. There must be assurance that not only the analytical technique is reliable, but also that the information system will perform to specification within the context of a regulated environment.

Further, with the increasing use of networks, the laboratory must be seen as part of the larger manufacturing complex. The trend towards geographically distributed databases emphasizes the need for the analysis and design of a corporate data architecture, a LIMS being a subset of this architecture. Full consideration must therefore be given to the (future) integration of databases and the formation of a corporate data architecture, thereby avoiding the dangers of having an 'island of automation'. Without doubt though, due to the advances in hardware, software, and networking standards, companies must define their corporate data architecture and the relationship between business strategy and information technology. Laboratory data systems must be designed within this context of a corporate information technology policy; for large multinational companies this would apply world-wide.

9.4 LABORATORY SYSTEM DEVELOPMENT LIFE CYCLE

Use of structured analysis and design and of the development life cycle helps ensure that the user requirements are fully specified and met. The underlying principles are: user involvement (the user must be involved in the development process from the start); quality assurance (quality assurance reviews ensure that the end product of each stage is fully considered for quality, completeness, consistency, and applicability); separation of logical and physical (hardware-/software-independent design enables the designer to consider the best way to satisfy user requirements before considering physical details); small, well-defined activities in a specified sequence; and diagrammatic modelling techniques.

There are various structured methods such as the Structured Systems Analysis and Design Method (SSADM), Jackson Structured Design (JSD), and Learmouth Systems Design Method (LSDM). Structured methods employ greater rigour than the standard SDLC method. Computer Aided Software Engineering tools are available for methods such as SSADM and these allow the automatic creation, maintenance, and checking of SSADM diagrams.

9.4.1 Initial Statement of Requirements

Objectives: to define user requirements;
                authorization to start investigation;
                formation of the project team
Inputs: various
Activities: feasibility study as appropriate
Output: approval to proceed to the next stage;
         terms of reference or start up document
Responsibility: senior management
Output destination: laboratory management

The initial statement of requirements is the identification of a strategic need and provides the authorization to proceed with the analysis. This statement is a broad description of the requirements and the associated goals and objectives. Where possible, these should be quantified. For very large projects, there may be a feasibility study in order to provide further details. At the end of the feasibility study there should be a cost/benefit analysis to provide justification, or not, for approval to proceed.

9.4.2 Stage 1—Investigation of the Current System

Objectives: to obtain a complete and accurate definition of the current
            system
Inputs: terms of reference document
Activities: initiate analysis;
           investigate the current system;
           complete the problems and requirements list;
           verification of stage 1
Output: a complete and documented description of the current system
         (*e.g.*, DFDs, LDSs);
         initial problems and requirements list;
         project task list and estimates;
         time schedule;
         project control and quality assurance mechanisms;
         approval to proceed to the next stage
Verification activities: review and assessment of the output of stage 1
Responsibility: project team
Output destination: project team

At this point there is a named project team consisting of representatives from the laboratory and systems development department as appropriate. In regulated environments, there must be representatives from the QAU on the team. This first stage is concerned with the analysis of current system operation and ongoing problems, and therefore must provide a documented, complete, and accurate understanding of the following: problems and requirements; current data flows and process operations; current data and data volumes; and boundaries of the system. The present system may be a manual system with card indexes, an obsolete computer system or a combination of the first two.

Stage 1, step 1—Initiate Analysis. The terms of reference (TOR) or start up document are reviewed to provide an overview and the scope of the analysis for sizing and partitioning. It is important to ensure that the TOR document reflects the current situation. It is important to note that the team must have a complete knowledge of the operations of the laboratory and also executive authority to make decisions. The scope is not only the system under investigation, but also other areas of the organization that may be affected. For large systems, project partitioning is essential. The running compilation of problems associated with the current system is called the problems and requirements list (PRL).

Stage 1, step 2—Investigate the Current System. The objective of stage 1, step 2 is to perform an initial overview analysis to provide a basis for estimation. For example, DFDs indicate the main data flows between the principal functional groups in the defined area under investigation. Logical data structures define the principal data structures. Both the high level DFDs and the initial LDSs provide a documentation framework.

The details of large, complex systems are controlled by levelled diagrams, which identify major documents, define the system boundaries, identify data flows, and identify processes. It is often useful to define explicitly the structural organization of personnel.

Stage 1, step 3—Problems and Requirements List. Users' demands and identified problems are placed on a prioritized 'wish list' in stage 1, step 3. This step formalizes these demands.

Stage 1, step 4—Verifications. Stage 1, step 4 is a quality assurance review to ensure that stage 1 has been completed to the team's satisfaction. A formal quality assurance meeting must take place. Project controls and quality assurance mechanism are established in accordance with agreed standards. The principle verification activities are the review and assessment of the analysis stage for completeness, consistency, and correctness. Hence a firm basis is provided for understanding the current system strengths and weaknesses prior to defining the requirements of the proposed system and clarifying the users' perceptions of problems and requirements.

9.4.3 Stage 2—Detailed Specification

Objectives: to produce a detailed design specification
Inputs: a complete documented description of the current system (*e.g.*,
            DFDs, LDSs);
            initial problems and requirements list;

            project task list and estimates;
            time schedule;
            project control and quality assurance mechanisms
Activities: define the logical system;
            define the security, audit, and control requirements;
            define and consolidate user requirements;
            identify System Operations (SOs);
            select SOs;
            verification of stage 2
Output: logical system definition;
            security and access requirements;
            prioritized PRL;
            selected SOs;
            approval to proceed to the next stage
Verification activities: review and assessment of the output of stage 2
Responsibility: project team
Output destination: project team

Stage 2 involves examination of the current data flows, data, and PRLs previously identified in order to produce a specification for the new system. A detailed specification of the required system is essentially a logical view which provides all the information about a system but is implementation independent. This stage is also concerned with the consolidation of user requirements as expressed in the PRL in the context of audit, control, and security requirements.

Stage 2, step 1—Define the Logical System. Logically all data should be stored once with all processes able to access this data as appropriate (*i.e.*, a database). This step involves the conversion of the current physical system DFDs into a logical view—logicalization. This is achieved by eliminating external physical constraints, removing duplication (of processes and data stores), and removing redundancy (of processes and data stores).

This step also helps to further identify problems with the current system. For example, duplication of process means that resources are being wasted in unnecessary effort. Duplication of data leads to inconsistency.

Stage 2, step 2—Define Security, Audit, and Control Requirements. It is important to address the security, audit, and control requirements early in the system life cycle. This step considers the issues and the output is a security, audit, and control requirements list. The main tasks are: identify the requirements for restricted access (privacy) on a 'need to know' basis. Considerations include: passwords, data access rights, and data update rights; identify audit requirements; identify exception handling criteria, *e.g.*, criteria for resubmission of exceptions; identify fallback and recovery requirements. Considerations include: back up procedures, recovery times, and safe storage of primary data; and identify performance criteria.

Stage 2, step 3—Define and Consolidate User Requirements. Step 3 involves a review of the PRL so far. Meetings held with relevant users determine the priority of each requirement, *i.e.*, essential or desirable. For example, the improvement of sample registration tracking and control along with data storage and security are considered essential requirements in order to meet the GLP regulations; potential benefits of the requirement, for example, the improved sample management leading to laboratory accreditation; resolution of conflicting requirements, for example, prioritization of sample analyses in the context of external fee paying clients and internal non-fee paying clients; and organizational implications—the most obvious being direct data entry by the analyst and the associated reduced work load for the technicians. It should be noted that some criteria, though important, are not always directly quantifiable, such as quality of service. As a consequence an agreed document can be written defining: organization and staffing possibilities; physical implementation constraints; and time/cost constraints.

Stage 2, step 4—Identify and Select from System Options. The system options (SOs) are not based on specific hardware and software solutions—such decisions are made later. Rather, the organization and scope of the system are examined and the effects on the laboratory are considered. The possible solutions take the form of DFDs (Level 1) supported by explanatory text. When developing a computer system there is no 'right solution'. The two extremes are: satisfying the entire PRL at a high cost and resolving immediate problems while postponing other problems.

The purpose of this step is to produce a menu of SOs allowing users to decide what they want and what they will pay for. Each option will differ in terms of functionality and impact on the organization with an associated cost/benefit profile. Up to 6 SOs may be produced with cost/benefit profiles. A more detailed specification is then produced for the chosen option. For example, for a multi-site laboratory all the major activities could be centralized or distributed. Furthermore, there are different degrees of automation that may be introduced.

Stage 2, step 5—Verification. Stage 2, step 5 is a quality assurance review to ensure that stage 2 has been completed to the team's satisfaction. A formal quality assurance meeting must take place.

9.4.4 Stage 3—Selection of Technical Options

Objectives: identification and selection of a suitable technical option
Inputs: logical system definition;
      security and access requirements;
      prioritized PRL;
      selected SOs
Activities: create technical options;
      selection of technical options;
      completion of required system specification;
      define design objectives;
      verification of stage 3
Output: menu of technical options;
      selected technical option;
      required system specification;

required system design objectives;
approval to proceed to the next stage
Verification activities: review and assessment of the output of stage 3
Responsibility: project team
Output destination: project team

Stage 3 identifies and specifies the possible ways in which the required system may be implemented. It is necessary to make the major decisions about the final implementation before the detailed design stage—the physical implementation can fundamentally affect the design at the logical level.

Stage 3, step 1—Create Technical Options. Stage 3, step 1 involves the creation of a menu of options which describe possible physical implementations of the required SOs that address the prioritized PRL. The constraints of the system base are defined. These features are imperative requirements of the system and include items such as cost. With a fixed budget none of the options must exceed the budget limit. It is often difficult to give exact costings, however, the work done in the earlier stages provides a good basis for sizing the system and thus the type of equipment needed. If some of the functions are particularly costly it is often useful to have separate cost/benefits and timescales. There may be pressing and mandatory requirements for a quick solution; physical implementation constraints; organizational constraints; and staffing constraints.

A menu of outline options based on possible technical approaches that satisfy the base constraints is created. The options should address: technical environment description which only needs to include the type, quality, and distribution of the equipment; the functional description which for each option indicate differences in processing mixtures (batch/on-line, *etc.*); the data communications requirements; the impact analysis, *i.e.*, the effects that the implementation of the option will produce in the user environment, in terms of the organizational and operational changes; the development and implementation plan which will be an outline plan indicating the resources and timescale needed for the development and implementation of the option; and a cost/benefit analysis to provide an objective yardstick on which to compare options.

The possible solutions to these options include in-house system development, turnkey system development, or the use of a supported commercial application package. Depending on the circumstances, particularly turnkey procurement, it may not be possible to define the hardware/software environment. In cases such as these it is necessary to limit the description of the technical environment to such factors as the location of terminals and performance requirements.

Vendor viability must be determined to ensure product reliability and longevity. Considerations include: how long has the vendor been established?; how big is the vendor, *i.e.*, number of staff?; does the vendor comply with international standards for the production of software, *i.e.*, BS and ISO standards?; what support is available?; what other clients have been supplied?; are previous clients satisfied with the product?; is the product continually being developed?; and availability of product upgrades? This information is invariably not available to end users but copies are deposited with recognized agents such as ESCROW, the

National Computing Centre (NCC). New releases are deposited in the ESCROW account and available for scrutiny by the regulatory agencies.

Stage 3, step 2—User Selects from the Technical Options. In this step the client is presented with the technical options for evaluation and selection.

Stage 3, step 3—Complete and Review the Required System Specification. This step 3 provides the basis for the design phase. It involves incorporating the effects of the decisions taken in selecting the chosen option into the required system specification produced in stage 2. Considerations include the performance criteria which the implemented system must achieve and which will provide performance targets for the design stage. This is the final step in the analysis phase. The main tasks are concerned with identifying data volumes. Considerations include: database/file sizes, main memory requirements, secondary memory requirements, and specifying function timing criteria.

Depending on the applications, the demands on the system may be 'peaky'. It is necessary to consider highest interactive peaks, most critical on-line response, and largest batch volumes. On-line transaction processing, that we have previously met, is a useful benchmark for evaluating computer systems and, as the name implies, takes into account on-line data processing requirements.

It is important to specify recovery criteria such as mean time between failures, maximum time to restore system, or performance level of back-up system.

Stage 3, step 4—Verification. Stage 3, step 4 is a quality assurance review to ensure that stage 3 has been completed to the team's satisfaction. A formal quality assurance meeting must take place.

9.4.5 Stage 4—System Design

Objectives: to produce the new computer system
Inputs: selected technical option;
             required system specification;
             required system design objectives
Activities: design;
               programming
Output: complete, fully documented system design;
            system code and/or database (LIMS);
            instrumentation interfaces;
            system test plan;
            approval to proceed to the next stage
Verification activities: review and assessment of the output of stage 4
Responsibility: systems analysts/designers;
                    programmers;
                    project team
Output destination: project team

Stage 4 is for defining the data structures, modules, interfaces, *etc.* as required. If a LIMS is used this stage would be concerned with customizing the software and writing the instrumentation interface software. Software design can be taken to include the conversion of the required system specification into a target language or database implementation

Stage 4, step 1—Design. Here we are concerned with the design of modules each with a set of specifications containing all the information necessary to write programs. This is followed by the conversion of the high level design modules to target codes. Each module that is coded must have a test plan.

Stage 4, step 2—Test Plan. Testing is the process of exercising or evaluating a system or system component by manual or automated means to verify that it satisfies specified requirements or to identify differences between expected and actual results (**FIPS Pub 101**, June 1983, ANSI/IEEE Std 729, 1983).

System testing is the process of testing an integrated hardware and software system to verify that the system meets its requirements (**FIPS Pub 132**, May 1988, ANSI/IEEE Std 1012, 1986 and Std 729, 1983).

A test plan is a document describing the scope, approach, resources, and schedule of intended testing activities. It identifies test items, the features to be tested, the testing tasks, who will do each task, and any risks requiring contingency planning (**FIPS Pub 132**, May 1988, ANSI/IEEE Std 1012, 1986 and Std 729, 1983).

The purpose of this step is to define the test plan, system testing procedures, responsibilities, test data, and expected test results. The main tasks are to create a test plan for each module, a system integration test plan (to test the system for consistency), a requirements test plan (for testing the system against the requirements), an acceptance test plan (for testing to be carried out by the user), a volume test plan (for the volume testing of the system), and a test phase plan. It should be noted that functional testing can be defined as 'the application of test data derived from the specified functional requirements without regard to the final program structure' (**FIPS Pub 101**, June 1983) and as such is designed to determine any discrepancies between expected and actual performance.

Stage 4, step 3—Verification. Stage 4, step 3 is a quality assurance review to ensure that stage 4 has been completed to the team's satisfaction. A formal quality assurance meeting must take place.

9.4.6 Stage 5—Integration and Testing

Objectives: to produce a complete, fully documented, integrated, and tested system

Inputs: complete, fully documented system design;
system code and/or database (**LIMS**);
instrumentation interfaces;
system test plan

Activities: test each module in the system;
test module integration;
test the interfaces to instruments;
test the complete computer system;
test the total system (manual and automatic functions)

Output: complete, fully documented system design;
        tested system code and/or database (LIMS);
        tested instrumentation interfaces;
        tested complete computer system;
        tested total system;
        system handover plan;
        approval to proceed to the next stage
Verification activities: review and assessment of the output of stage 5
Responsibility: systems analysts/designers;
                 programmers;
                 project team
Output destination: project team

The output of stage 5 will be documented, integrated, and tested software. Stage 5 is concerned with the complete system. For very large systems this is a major consideration.

Stage 5, step 2—Handover Planning. When the new system has been completely tested, *i.e.*, the files or database created and the hardware installed, there must be a handover from the old to the new system. This handover, also called changeover, must be managed. There is no universal 'best' method, various approaches include:

> Direct handover or 'Big Bang'—with this method the old system ceases abruptly and is replaced immediately by the new system. The handover is best done at the weekends or during a vacation period when there is an obvious natural break. This direct method is efficient in that it minimizes work duplication but demands very careful testing, planning, and attention to operational detail. The user must have complete confidence that the new system will work first time, every time. The main disadvantage is the possibility that the new system is not fault free. This problem is exacerbated if there are no results from the old system to compare with those from the new once handover has occurred. Furthermore, if the new results are incorrect, it is difficult to make corrections to the system and at the same time keep it fully operational.
>
> Parallel running—with this method the old and new systems are run concurrently for a period of time. This means that the two sets of results and operational characteristics can be immediately compared. Should there be problems with the new system, the old system continues in use until the necessary corrective action has been taken. The major disadvantage is the duplication in effort.
>
> Phased handover—this is similar to parallel running except that at the start only a subset of the data is run in parallel, *e.g.*, sample logging. Should no problems arise, this subset increases up to one parallel run. Advocates of this method claim there is less duplication of effort than with parallel running.
>
> Pilot running—The new system is employed to re-run the source data from a previous period.

Even with relatively small systems, there is considerable effort involved in handover. The two main methods for activity control are bar charts and network analysis. These methods are also applicable to other activities such as systems analysis.

Bar charts, also known as Gantt charts, consist of a list of activities marked off in time periods. Each activity is entered initially as a horizontal line denoting its duration and scheduled periods. When progress is made, these lines (bars) are noted to indicate the status to date. The major disadvantages are that this method does not fully represent the interdependence of activities or facilitiate the planning of resources and costings.

Network analysis, also called 'critical path analysis/method', is suited to the management of large numbers of activities. In network analysis for handover planning, a list of activities is compiled together with the estimated times, resources needed, and preceding activities. From this information a network diagram is constructed according to the following rules: each activity is drawn as an arrowed line and is preceded by the activities that must be completed prior to its commencement; no activity is preceded by an activity that does not have to be completed before it; and an event circle is interposed between consecutive activities. Though the rules are simple, considerable care must be taken to ensure the correctness of the diagram.

Stage 5, step 3—Verification. Stage 5, step 3 is a quality assurance review to ensure that stage 5 has been completed to the team's satisfaction. A formal quality assurance meeting must take place.

9.4.7 Stage 6—Implementation and Handover

Objectives: deliver a fully documented, tested, and validated system to users

Inputs: complete, fully documented system design;
tested system code and/or database (LIMS);
tested instrumentation interfaces;
tested computer system;
system handover plan

Activities: manage an agreed handover strategy

Output: fully validated and operational system;
operating instructions/manual procedures;
education and training plan

Verification and validation activities: review and assessment of the output of stage 6;
validation according to statement of requirements

Responsibility: systems analysts/designers;
programmers;
project team

Output destination: users

Validation is the process of evaluating a systems or system software at the end of the system or software development process to assure compliance with user requirements. As such it ensures a laboratory computer system that will be acceptable to the regulatory authorities. In the final analysis, validation is the responsibility of the end user. Validation occurs according to a validation plan at the end of the system development and prior to handover. It must occur irrespective of whether the system was developed in-house, under contract, or vendor supplied. Acceptance testing is done to determine whether or not a system satisfies its acceptance criteria and to enable the customer to determine whether or not to accept the system.

Stage 6, step 1—Create Operating Instructions. In this step the operating instructions (SOPs) are defined. The actual content of the instructions will depend on the installation standards. This will take the form of a User Manual. An essential component of handover is an education and training program for end users. This helps for the more efficient use of the new system.

Stage 6, step 2—Validation. The validation plan must provide a clear description of the process by which a particular system is validated.

Validation must occur regardless of whether the system was developed in-house or was vendor supplied and is the responsibility of the end user. The validation plan is also called the acceptance test plan or the validation protocol. It must provide a clear, unambiguous description of the procedures and their chronological order by which the system will be validated. As such the plan must include a description of the system (environment and functions) and the responsibilities of individuals. In effect the plan must be a complete document that could be used by a party to reproduce the validation of the given system. The documentation of the development life cycle is invaluable. Hence validation plans typically have sections that include: system description, test operations to be performed, assumptions made, limitations, items not tested, responsibilities, test data and expected results, acceptance criteria, and error resolution.

After the successful execution of the validation plan, the system is ready for approval by the senior management, the QAU, and users.

9.4.8 Stage 7—Maintenance

> Objectives: to maintain the system in a validated state
> Inputs: perfective maintenance requests;
>   adaptive maintenance requests;
>   corrective maintenance requests
> Activities: documented change control and revalidation
> Output: documented change control and revalidation;
>   system in validated state
> Verification and validation activities: as appropriate
> Responsibility: systems analysts/designers;
>   programmers;
>   project team;

users;

QAU

Computer systems are dynamic. For any system, with time, there will be changes in hardware, software, standard operating procedures, *etc.*, consequently, there must be change control. Change control is the on-going evaluation of system operations and maintenance during the use of a system to determine when and if repetition of a validation process or a specific portion of it is needed in order to maintain a product in a validated state. The important criterion is that the user must be assured that, even after the changes, the system is in a validated state. The amount of testing will depend on what changes are made. For example, cosmetic changes such as error diagnostic messages require minimal verification. The test should use the original test data to demonstrate that the changes introduced have not affected the performance of the system. Any differences in the test results should be taken as being symptomatic, and must be documented, investigated, and resolved. It is important to note that change control procedures apply irrespective of the origin of the change.

In recognition of the problems associated with large real-time systems Software Tools for Application to large Real Time Systems (STARTS) was published in 1986 by the National Computing Centre (NCC), UK. The guide contains a chapter on the project life cycle (Figure 9.5) and outlines the 'best

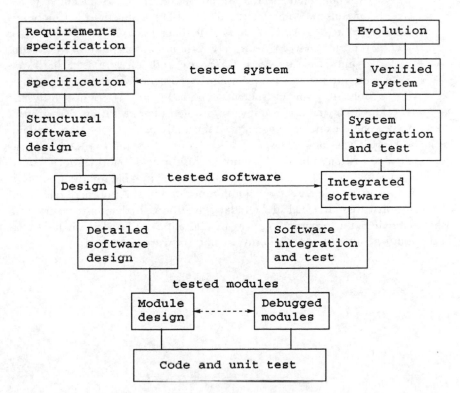

Figure 9.5

practices' employed in large real-time systems. Certainly a major problem in any very large system is project control and management. Projects in controlled environments (PRINCE) is a system that comprehensively addresses the management, technical, and quality assurance tasks in project management. PRINCE was developed by the NCC in conjunction with Praxis Systems. As such, it is fully compatible with most structured design methods such as SSADM, JSD, and LSDM.

9.5 INSPECTION

Inspections of computer-based systems in, for example, the pharmaceutical industry are the responsibility of the MCA in the UK and the FDA in the US. The inspectorates are concerned with all computer systems that affect product quality. Within the UK, the inspectorate has reported considerable problems including the following: suppliers not being audited to ensure that the software produced is of acceptable quality; no current system documentation (*e.g.*, DFDs); poor disaster recovery procedures; incorrect specification of computer hardware disk capacity; no documentation for SDLC; responsibilities for system control not defined; lack of formal protocols for revalidation; and no user-training records.

The problems reported in the US are no better. In 'GMP Documentation Requirements for Automated Systems: Part III, FDA Inspections of Computerised Laboratory Systems', *Pharm. Tech. Int.*, October 1992, Tezlaff, an inspector with the FDA reported that 'FDA inspections of laboratory automated systems may be divided into two categories—hardware and software. The hardware consists of the computer (CPU, floppy or hard drives, modems, *etc.*), the interface boxes, the wiring, and the instrumentation or apparatus connected to the computer'. The following are examples of problems found during FDA inspections: 'Accuracy—the analogue signal from the analyser to the computer had not been checked for accuracy'; 'Frequently, FDA inspectors find that such applications (software written in house) have not been subjected to formal documented validation testing'; and 'Parallel testing can provide valuable information but is unacceptable as the sole basis for validation because it is not normally designed to test the software source code at boundary conditions, to demonstrate program behaviour when the program receives invalid inputs . . .'. In conclusion, as Tezlaff put it: 'Actual testing of hardware and software is necessary'.

Bibliography

Chapter 1
Atkin, 'Computer Science',
Watkin, 'Computer Technology for Technicians and Technician Engineers',

Chapter 2
East, 'Computer Architecture and Organisation',
Tanenbaum, 'Structured Computer Organisation',
Preperata, 'Introduction to Computer Engineering',
Zaks and Wolfe, 'From Chips to Systems',

Chapter 3
As for Chapter 2

Chapter 4
Peterson and Silberschatz, 'Operating System Concepts'
Tanenbaum, 'Operating Systems, Design and Implementation',

Chapter 5
Cripps, 'Computer Interfacing—Connection to the Real World',
Campbell, 'The RS232 Solution—How to Use Your Serial Port',
Wilkinson and Horrocks, 'Computer Peripherals',

Chapter 6
Halsal, 'Data Communications, Computer Networks, and OSI',
Tanenbaum, 'Computer Networks',
Freer, 'Computer Communications and Networks',

Chapter 7
Hawryszkiewycz, 'Introduction of Systems Analysis and Design',
Hanson, 'Design of Computer Data Files',
        'Language Independent Design—An Introduction', NCC/Blackwell
Hanson, 'Essentials of Computer Data Files',
E.F. Codd, 'A Relational Model of Data for Large Shared Systems', 1970

Chapter 8

Pressman, 'Software Engineering, a Practitioner's Approach',

Burgess, 'Structured Program Design',

Maj, 'Language Independent Design—An Introduction',

'OECD Test Guide-lines', 1981,

'OECD Principles of Good Laboratory Practice', 1981,

'Compliance Policy Guide #7132a.07', FDA, October 1982, Chapter 32a: Computerised Drug Processing: Input/Output Checking;

'Compliance Policy Guide #7132a.08', FDA, December 1982, Chapter 32a: Computerised Drug Processing: Identification of Persons on Batch and Control Records;

'Guide to Inspection of Computerised Systems in Drug Processing', FDA, February 1983 ('Blue Book');

'Compliance Policy Guide #7132a.11', FDA, December 1984, Chapter 32a: Computerised Drug Processing: CGMP Applicability to Hardware and Software;

'Compliance Policy Guide #7132a.12', FDA, January 1985, Chapter 32a: Computerised Drug Processing: Vendor responsibility;

'Compliance Policy Guide #7132a.15', FDA, April 1987, Chapter 32a: Computerised Drug Processing: Source Code for Process Control Application Programs; and

'Software Development Activities. Reference Materials and Training Aids for Investigators—Technical Report', FDA, July 1987.

Chapter 9

'Analysis and design of automated laboratories in regulated environments: Good Laboratory and Clinical Practices—Techniques of the Quality Assurance Professional', eds. Carson and Dent,

Avison and Fitzgerald, 'Information Systems Development',

Nicholls, 'Introducing SSADM, the NCC guide',

'Good laboratory practice, the UK compliance program', Department of Health, London,

Good laboratory practice, the application of GLP principles to computer systems', Department of Health, London,

'Computerised data systems for nonclinical safety assessment, current concepts and quality assurance', Drug Information Association, (also known as the Red Apple book)

'STARTS Purchasers Handbook', NCC, 1986

'Computerised systems and GMP, a series of articles by A.J. Trill (MCA)', *Pharm. Tech. Int.*

Subject Index and Abbreviations